# SLICING THE TRUTH

## On the Computable and Reverse Mathematics of Combinatorial Principles

# LECTURE NOTES SERIES
## Institute for Mathematical Sciences, National University of Singapore

Series Editors:  Chitat Chong and Wing Keung To
*Institute for Mathematical Sciences*
*National University of Singapore*

ISSN: 1793-0758

---

*Published*

*For the complete list of titles in this series, please go to
http://www.worldscientific.com/series/LNIMSNUS

Lecture Notes Series, Institute for Mathematical Sciences,
National University of Singapore

Vol.
28

# SLICING THE TRUTH

## On the Computable and Reverse Mathematics of Combinatorial Principles

Denis R Hirschfeldt

The University of Chicago, USA

Editors

Chitat Chong
National University of Singapore, Singapore

Qi Feng
Chinese Academy of Sciences, China

Theodore A Slaman
University of California, Berkeley, USA

W Hugh Woodin
Harvard University, USA

Yue Yang
National University of Singapore, Singapore

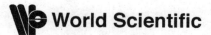 World Scientific

NEW JERSEY · LONDON · SINGAPORE · BEIJING · SHANGHAI · HONG KONG · TAIPEI · CHENNAI

*Published by*

World Scientific Publishing Co. Pte. Ltd.
5 Toh Tuck Link, Singapore 596224
*USA office:* 27 Warren Street, Suite 401-402, Hackensack, NJ 07601
*UK office:* 57 Shelton Street, Covent Garden, London WC2H 9HE

**Library of Congress Cataloging-in-Publication Data**
Hirschfeldt, Denis Roman, author.
    Slicing the truth : on the computable and reverse mathematics of combinatorial principles /
Denis R. Hirschfeldt, The University of Chicago, USA ; editors, Chitat Chong, National University
of Singapore, Singapore, Qi Feng, Chinese Academy of Sciences, China, Theodore A. Slaman,
University of California, Berkeley, USA, W. Hugh Woodin, Harvard University, USA, Yue Yang,
National University of Singapore, Singapore.
    pages cm -- (Lecture notes series / Institute for Mathematical Sciences, National University of
Singapore ; vol. 28)
    Includes bibliographical references and index.
    ISBN 978-9814612616 (hardcover : alk. paper) -- ISBN 9814612618 (hardcover : alk. paper)
    1. Reverse mathematics. 2. Combinatorial analysis. I. Chong, C.-T. (Chi-Tat), 1949–   editor.
II. Title.
    QA9.25.H57 2015
    511'.6--dc23

                                                                              2014027104

**British Library Cataloguing-in-Publication Data**
A catalogue record for this book is available from the British Library.

Printed in Singapore

# Contents

# Foreword
# by Series Editors

The Institute for Mathematical Sciences (IMS) at the National University of Singapore was established on 1 July 2000. Its mission is to foster mathematical research, both fundamental and multidisciplinary, particularly research that links mathematics to other efforts of human endeavor, and to nurture the growth of mathematical talent and expertise in research scientists, as well as to serve as a platform for research interaction between scientists in Singapore and the international scientific community.

The Institute organizes thematic programs of longer duration and mathematical activities including workshops and public lectures. The program or workshop themes are selected from among areas at the forefront of current research in the mathematical sciences and their applications.

Each volume of the *IMS Lecture Notes Series* is a compendium of papers based on lectures or tutorials delivered at a program/workshop. It brings to the international research community original results or expository articles on a subject of current interest. These volumes also serve as a record of activities that took place at the IMS.

We hope that through the regular publication of these *Lecture Notes* the Institute will achieve, in part, its objective of reaching out to the community of scholars in the promotion of research in the mathematical sciences.

July 2014

Chitat Chong
Wing Keung To
*Series Editors*

# Foreword
# by Volume Editors

The series of Asian Initiative for Infinity (AII) Graduate Logic Summer School was held annually from 2010 to 2012. The lecturers were Moti Gitik, Denis Hirschfeldt and Menachem Magidor in 2010, Richard Shore, Theodore A. Slaman, John Steel, and W. Hugh Woodin in 2011, and Ilijas Farah, Ronald Jensen, Gerald E. Sacks and Stevo Todorcevic in 2012. In all, more than 150 graduate students from Asia, Europe and North America attended the summer schools. In addition, two postdoctoral fellows were appointed during each of the three summer schools. These volumes of lecture notes serve as a record of the AII activities that took place during this period.

The AII summer schools was funded by a grant from the John Templeton Foundation and partially supported by the National University of Singapore. Their generosity is gratefully acknowledged.

July 2014

Chitat Chong
National University of Singapore, Singapore

Qi Feng
Chinese Academy of Sciences, China

Theodore A. Slaman
University of California, Berkeley, USA

W. Hugh Woodin
Harvard University, USA

Yue Yang
National University of Singapore, Singapore

*Volume Editors*

# Preface

When Rod Downey and I finished our book *Algorithmic Randomness and Complexity*, which was almost a decade in the making, I promised myself I would never again write a book. But accidents happen. In 2010, I was invited to give a short course at the Asian Initiative for Infinity Graduate Summer School, organized by the Institute for Mathematical Sciences and the Department of Mathematics of the National University of Singapore, and to write a version of my lecture notes for publication. The topic of the course was the reverse mathematics and computability theory of combinatorial principles, an area of research whose roots reach back several decades, but which has seen a particular surge of activity in the last few years. Much of this work has proceeded along lines that are fairly distinct from the material covered in Simpson's excellent *Subsystems of Second Order Arithmetic*, and there has been little alternative to reading research articles for those interested in understanding it. While reading original papers is highly recommended, it can be a difficult process without appropriate guidance. I wanted my notes to be an entryway into this area, providing both an overview of some fundamental ideas and techniques, and enough context to make it possible for students with at least a basic knowledge of computability theory and proof theory to appreciate the exciting advances currently happening in the area, and perhaps make contributions of their own.

I decided to adopt a case-study approach, using the study of versions of Ramsey's Theorem (for colorings of tuples of natural numbers) and related principles as illustrations of various aspects of computability theoretic and reverse mathematical analysis. Even within this deliberately narrow focus, I felt no need to be encyclopedic. It was not my goal to write a survey, but to tell a story. Nevertheless, when it comes to mathematics, a properly illustrative story needs details. Furthermore, thorough discussions of some of these details were difficult to find in existing sources. And while there may be some who can tell a long story without digressions, I am not among them. Thus the text grew and grew, until it became a book. So be it.

I will give an overview of the book in Chapter 1, but for now, here is the abstract I wrote when still thinking of this text as an article: We discuss two closely related approaches to studying the relative strength of mathematical principles, computable mathematics and reverse mathematics. Drawing our examples from combinatorics and model theory, we explore a variety of phenomena and techniques in these areas. We begin with variations on König's Lemma, and give an introduction to reverse mathematics and related parts of computability theory. We then focus on Ramsey's Theorem as a case study in the computability theoretic and reverse mathematical analysis of combinatorial principles. We study Ramsey's Theorem for Pairs ($\mathrm{RT}^2_2$) in detail, focusing on fundamental tools such as stability, cohesiveness, and Mathias forcing; and on combinatorial and model theoretic consequences of $\mathrm{RT}^2_2$. We also discuss the important theme of conservativity results. In the final section, we explore several topics that reveal various aspects of computable mathematics and reverse mathematics. An appendix contains a proof of Liu's recent result that $\mathrm{RT}^2_2$ does not imply Weak König's Lemma. There are exercises and open questions throughout.

# Acknowledgments

I was partially supported during the writing of this book by grants DMS-0801033 and DMS-1101458 from the National Science Foundation of the United States. This book is a version of a short course given at the Asian Initiative for Infinity Graduate Summer School, sponsored by the Institute for Mathematical Sciences and the Department of Mathematics of the National University of Singapore from 28 June to 23 July, 2010, and funded by the John Templeton Foundation and NUS. I thank these organizations; the organizers Ted Slaman and Hugh Woodin; our hosts at NUS Chi Tat Chong, Qi Feng, Frank Stephan, and Yue Yang; the other lecturers Moti Gitik and Menachem Magidor; and all of the participants for a delightful and rewarding experience. I also thank the Einstein Institute of Mathematics of The Hebrew University of Jerusalem for hosting a visit during which much of this book was written, Menachem Magidor for arranging this visit, and the students in a short course I taught there based on a draft version of this book. Finally, I thank Tsvi Benson-Tilsen, Chi Tat Chong, Damir Dzhafarov, Bill Gasarch, Noam Greenberg, Jeff Hirst, Carl Jockusch, Joe Mileti, Joe Miller, Antonio Montalbán, Ludovic Patey, Ted Slaman, Reed Solomon, Wei Wang, and Yue Yang for useful comments and responses to queries.

# Chapter 1

# Setting Off: An Introduction

Every mathematician knows that if $2 + 2 = 5$ then Bertrand Russell is the pope. Indeed, Russell is credited with having given a proof of that fact in a lecture by arguing as follows: If $2+2 = 5$, then, subtracting 3 from each side, $1 = 2$. The pope and Russell are two, therefore they are one. Of course, from the point of view of classical logic, no such proof is needed, since a false statement implies every statement. Contrapositively, every statement implies a given true statement. But suppose we were to take seriously the task of proving that, say, the Four Color Theorem implies that there are infinitely many primes. What are the chances that any of us could come up with a proof that "really uses" the Four Color Theorem? The exercise may seem as pointless as it is difficult, but of course mathematicians do set and perform tasks of this kind on a regular basis. "Use the Bolzano-Weierstrass Theorem to show that if $f : [0,1] \to \mathbb{R}$ is continuous, then $f$ is uniformly continuous." is a typical homework problem in analysis, and the question "Can Chaitin's information-theoretic version of Gödel's First Incompleteness Theorem be used to prove Gödel's Second Incompleteness Theorem?" led to a lovely recent paper by Kritchman and Raz [116]. There is also a well-established practice of showing that a given theorem can be proved *without* using certain methods, for instance in the exercise of proving the irrationality of $\sqrt{2}$ without using the fundamental theorem of arithmetic, or in elementary proofs of the prime number theorem. We have all heard our teachers and colleagues say things like "Theorems $A$ and $B$ are equivalent." or "Theorem $C$ does not just follow from Theorem $D$." or "Using Lemma $E$ in proving Theorem $F$ is convenient but not necessary." These are often crucial things to understand about an area of mathematics.

They are also things that can help us make connections between different areas of mathematics. For example, consider the following theorems: the

1

existence of suprema for continuous real-valued functions on $[0, 1]$, the local existence theorem for solutions of ordinary differential equations, Gödel's completeness theorem, the existence of primes ideals for countable commutative rings, and Brouwer's Fixed Point Theorem. Dissimilar as these theorems might seem, at heart they all involve compactness arguments in an essential way, and can all be seen as reflections in different fields of the same fundamental combinatorial idea, expressed in a principle known as Weak König's Lemma that we will discuss in some detail below. We will be able to make this claim formal in Section 4.4.

In this book, we will discuss two closely related approaches to making mathematically precise sense of this idea of establishing implications and nonimplications between provably true principles: computable mathematics and reverse mathematics. We will focus on combinatorial principles that are easy to state and understand, but exhibit intricate and intriguing behavior from these points of view. This book is not meant as a survey of results in this area, but rather as an introduction to a constellation of ideas and methods, unapologetically biased towards my own interests (particularly the computability theoretic and reverse mathematical analysis of combinatorial and model theoretic principles related to Ramsey's Theorem for pairs), but hopefully with enough breadth and depth to engage and motivate newcomers to the area. In particular, although the program of reverse mathematics has close ties with the foundations of mathematics, I will not say much about that aspect of the field.

I will assume some background in mathematical logic, in particular the basics of computability theory, though a few essential computability theoretic concepts will be reviewed briefly in Section 2.1. Otherwise, this book should be self-contained. There are exercises scattered throughout; working them out is an integral part of using this text. A few open questions will also be mentioned, and readers are encouraged to do battle with them as well. One never knows when a clever idea will solve a long-standing problem.

## 1.1 A measure of motivation

There are many things that comparing the relative strength of theorems can do for us. The process of revealing the "combinatorial core" of a theorem can give us significant insight. For example, it can tell us when a method is not just useful in proving a theorem, but in fact *necessary*. In other cases, it can suggest different ways to prove a theorem, or clarify the relationships between variations of a theorem. There are also foundational issues it can address, by pinpointing exactly how much of the abstract machinery of set theory is necessary to prove a given theorem, or a collection of basic theorems in a given area of mathematics. In giving examples of how such considerations can be of interest to mathematical practice (even outside the confines of mathematical logic), I could do no better than the ones outlined in Section 2 of Shore [183], so I will refer readers to that compelling discussion.

But even granted that questions of implication between mathematical principles are of interest, it is still reasonable to ask why we should attempt to study formal analogs of these questions, and, if we should, what would count as reasonable formalizations. I will not try to give a general answer to the second question. For one thing, there are many contexts other than the ones we will explore in which mathematical logic and theoretical computer science study the relative strength of theorems and constructions, from complexity theory to the study of the relationships between principles not provable in ZFC such as large cardinal axioms. Our framework sits somewhere in between, where we work "up to computable procedures" but consider set existence axioms of various strengths, all much below the full power of ZFC. Although one can argue abstractly for the appropriateness of particular choices of formal methods, in the end, I think it is only through developing the consequences of these choices that a real argument for their adequacy can be made, and it is the purpose of this book to give a glimpse at how this development has proceeded in the particular cases of computable mathematics and reverse mathematics.

As to the first question, one answer is the usual argument for bringing mathematical tools to bear on an area of inquiry. The rigor of the mathematical method, together with its highly developed tools, can often uncover things that less formal methods cannot. Of course, the suitability of various areas of inquiry to formalization and mathematization varies a great deal, but certainly one should expect mathematics itself to be highly amenable to this process. Indeed, the development of metamathematics, the mathematical study of mathematics itself, has been one of the great stories of the intellectual history of the past century and a half or so.

A particular aspect that formalization seems likely to help with is the question of how one argues that a certain principle $A$ does *not* imply another principle $B$. In some cases one can have informal plausibility arguments, or even more formal model theoretic ones, as when one can point out that versions of $A$ hold for a wider class of objects than corresponding ones of $B$. However, these methods are ad hoc and not generally available. As we will see, a formal approach to studying the relative strength of theorems gives a much more systematic and widely applicable way to establish such nonimplications.

Furthermore, it is always worth keeping in mind that the objects of our metamathematical analysis are "one level up" from the objects of ordinary mathematics. That is, while an algebraist might study groups, we study *theorems about groups*. Binary trees are reasonably simple objects, and the fact that every infinite binary tree has an infinite path is a simple result. But that fact (known as Weak König's Lemma) as a mathematical object in its own right, is considerably more complicated and interesting, as we will see. Much of this complexity can be revealed only through a formal, mathematical analysis.

Of course, we should never disregard the fact that formalization usually entails losses as well as gains. Simplifications and compromises must be made, and aspects of the original problem left out of the picture. There are likely to be instances in which the answers given in our formal setting to the kinds of questions mentioned above are unsatisfying, but I believe that enough answers of genuine interest can be given to justify our methods.

In addition to this "practical" justification for the kind of metamathematical work we will discuss, there is also a philosophical story to be told. With the increasingly abstract methods being introduced into mathematics in the late 19th and early 20th centuries, a need for increased rigor was felt. Bombshells such as Russell's Paradox, however, threatened to destroy the very foundations on which this rigor was built. (The set theoretic viewpoint was one of the hallmarks of this new style of mathematics, and phenomena like Russell's Paradox put the very concept of set itself into question.) Hilbert's Program was an attempt to resolve this foundational crisis by establishing the consistency of the whole vast apparatus of modern mathematics using only the kinds of hopefully uncontroversial methods that much of the more concrete mathematics of previous centuries had employed. In particular, Hilbert spoke of "finitistic" methods. These were to be highly concrete, constructive ones. A good example is given by the simple combinatorial manipulations of finite strings drawn from a finite alphabet involved in the notion of formal deduction in first order logic.

Thus, the hope was to take the large system $S$ consisting of all generally accepted mathematical methods (nowadays, we might think of $S$ as ZFC, say) and to prove the consistency of $S$ while working in a weak system

$T \subset S$ consisting only of finitistically acceptable principles, such as those involving simple manipulations of strings. Mathematicians would then be able to sleep in peace, knowing that the consistency of $S$ is as sure as that of $T$. This hope was shattered by Gödel's Second Incompleteness Theorem, which showed in particular that not even $S$ itself, let alone any such $T$, is powerful enough to prove the consistency of $S$ (unless, of course, $S$ is actually inconsistent, in which case it proves everything).

But the ashes of Hilbert's Program have proved a fine fertilizer. Methods of mathematical logic that could have been merely tools to settle a single problem (albeit an exceptionally important one) could now become instruments of fine analysis. Instead of a simple division between unexceptionable methods and doubtful ones in need of justification, work in the foundations of mathematics has revealed subtle gradations, and metamathematical work has provided formal analogs and results about where various theorems, methods, and even whole areas of mathematics fall in this foundational universe. Reverse mathematics in particular has been tied to such concerns from its outset, and its classification of the strength of mathematical principles into various levels has implications for this kind of foundational work. Some discussion of these matters can be found in Simpson [187, 190, 191]; see in particular the table on page 43 of [191]. As the present book is meant as a tutorial on the mathematical practice of reverse mathematics and computable mathematics, and as my own interest in these subjects does not stem primarily from such foundational considerations, but rather from a desire to understand (at a purely mathematical level) some of the complex interactions between "ordinary" mathematics, combinatorial structure, and computability, I will not say more on this subject, except to comment on a line from Borges' "Fragmentos de un Evangelio apócrifo":

> "*Nada se edifica sobre la piedra, todo sobre la arena, pero nuestro deber es edificar como si fuera piedra la arena.*"
> ["Nothing is built on stone, all on sand, but our duty is to build as if the sand were stone."]

The work of Gödel and others has shown that mathematics, like everything else, is built on sand. As Borges reminds us, this fact should not keep us from building, and building boldly. However, it also behooves us to understand the nature of our sand.

We finish this section with an important remark: The approaches to analyzing the strength of theorems we will discuss here are tied to the

countably infinite. Finite structures are of course of great interest, but
complexity theoretic methods are usually better suited to their analysis
than computability theoretic ones. In the other direction, the application
of computability theoretic and reverse mathematical methods to essentially
uncountable mathematics is still in its infancy. (Here "essentially uncountable" is meant to exclude areas where uncountable objects have reasonable
countable approximations, such as countable dense subsets of separable
metric spaces.) For a discussion of various approaches to uncountable computable mathematics (and reverse mathematics), see [73].

Simpson [191] makes a distinction between "set-theoretic" and "ordinary", or "non-set-theoretic", mathematics in formulating what he calls
the main question of his book: "Which set existence axioms are needed
to prove the theorems of ordinary, non-set-theoretic mathematics?" In the
former camp he places set theory itself, and other branches such as point-
set topology and uncountable discrete mathematics, which arose from the
development of set theory and involve essentially uncountable structures.
In the latter, he places countable algebra, analysis, number theory, and
so on, areas in which objects are either countable or have countable approximations. As he puts it, "the set existence axioms which are needed
for set-theoretic mathematics are likely to be much stronger than those
which are needed for ordinary mathematics. Thus our broad set existence
question really consists of two subquestions which have little to do with
each other. Furthermore, while nobody doubts the importance of strong
set existence axioms in set theory itself and in set-theoretic mathematics
generally, the role of set existence axioms in ordinary mathematics is much
more problematical and interesting." Because of our focus on countable objects, "infinite" below will mean countably infinite unless otherwise stated.

## 1.2  Computable mathematics

Computability theory gives us many tools to calibrate the complexity of
mathematical principles. Particularly fundamental is the idea of a set of
natural numbers $Y$ being computable in another set $Z$, which means that
there is an algorithm that, on input $n$, decides whether $n \in Y$ while using $Z$
as an *oracle*. That is, the algorithm is allowed to ask as many questions as it
wants about whether certain particular numbers are in $Z$ (but only a finite
number of questions for each input, of course, since if an algorithm is to
terminate, it must do so in finite time). We can formalize this notion using

Turing machines with oracle tapes (see e.g. Soare [196, 197]; of course, we can also use any other equivalent formalism), and we say that $Y$ is *computable in* $Z$, or *computable relative to* $Z$, or $Z$-*computable*.

In this section, we focus on a class of theorems that includes most of the ones studied below. Before describing this class, we should clarify a couple of terms. A *first order object* is a natural number or an object that can be coded as a natural number. For example, we can code a finite sequence of natural numbers $(n_0, \ldots, n_{k-1})$ as $\prod_{i<k} p_i^{n_i+1}$, where $p_0 < p_1 < \cdots$ are the primes. A *second order object* is a set of natural numbers or an object that can be coded as a set of natural numbers. For example, a set $S$ of finite sequences of natural numbers, such as a tree, can be coded as the set of all $\prod_{i<k} p_i^{n_i+1}$ for $(n_0, \ldots, n_{k-1}) \in S$. (We will discuss coding in more detail in Section 4.1.) Recall that we are assuming countability of all the objects we discuss.

Let us consider true principles that can be expressed in the form "for all $X$ in the class $\mathcal{C}$, there is a $Y$ bearing relation $R$ to $X$", where $X$ and $Y$ are second order objects and $\mathcal{C}$ and $R$ can be defined without quantification over second order objects (so that our principles do not depend on the existence of any such objects other than $X$ and $Y$). Examples of theorems of this form abound: every commutative ring has a prime ideal, every vector space has a basis, every consistent theory has a model, and so on. Let us consider in particular Weak König's Lemma (WKL), which, as mentioned above, states that every infinite binary tree has an infinite path. (See Section 3.1 for formal definitions.) For a principle $P$ in the above form, say that $X$ is an *instance* of $P$ if $X \in \mathcal{C}$, and that $Y$ is a *solution* to $X$ if $Y$ bears relation $R$ to $X$ (which we write as $R(X, Y)$). For example, an instance of Weak König's Lemma is an infinite binary tree $T$, and a solution to $T$ is an infinite path on $T$. The idea is that we think of $P$ as guaranteeing the existence of solutions to the problem, "given $X \in \mathcal{C}$, find a $Y$ such that $R(X, Y)$."

It is then natural to ask how difficult it is to obtain a solution $Y$ from an instance $X$ of $P$. We can measure this difficulty in terms of the arithmetic hierarchy, the lowness/highness hierarchy, or any of a large number of notions of computability theoretic strength. (See Section 2.1 for computability theoretic terminology and notation.) In the case of WKL, for instance, there has been a long history of answers to this question. Kreisel [114] showed that WKL is not computably true by providing a computable instance of WKL with no computable solution. On the other hand, he also showed that every instance $X$ of WKL has a solution computable in $X'$ (the halting problem relativized to $X$). Shoenfield [179] improved this result to show that every instance $X$ of WKL has a solution that is strictly weaker than $X'$ (in the sense of Turing reducibility). In their celebrated Low Basis Theorem, Jockusch and Soare [105] showed that, in fact, every instance $X$ of WKL has a solution $Y$ such that $Y'$ (and even $(X \oplus Y)'$) is computable in $X'$.

These and other results on the complexity of WKL, which we will further discuss in Chapter 3, have proved exceptionally useful throughout computability theory and its applications. An important reason is that there are many mathematical constructions that can be thought of as finding infinite paths on infinite binary trees. We will give an example from mathematical logic, Lindenbaum's Lemma, later in this section. In Exercise 3.5, we will give one from algebra, namely finding a prime ideal of a given commutative ring (in this book, all rings have units), and we will mention others in Section 4.4. As it turns out, many of these examples can be turned around to show that the use of WKL is not just a convenient tool, but in fact *essential*, meaning that although these constructions, and the corresponding existence theorems, deal with different kinds of mathematical objects, they can be thought of as having the same fundamental combinatorial core, which is expressed by WKL. We will illustrate this idea when we discuss Lindenbaum's Lemma.

Of course, there are many principles that have different combinatorial cores. To begin with, there are many principles that, unlike WKL, are computably true. For example, every field $F$ has an algebraic closure computable in $F$. Another interesting example is WKL restricted to trees with no dead ends (i.e., trees where every node has at least one successor). Given any infinite tree with no dead ends, we can easily compute an infinite path: at each step, just take the leftmost available immediate successor. In the opposite direction, there are many problems that are harder to solve than WKL. For example, for every $X$ there is a commutative ring $R$ that is computable in $X$, such that any maximal ideal computes $X'$. (Another example, as we will see in Chapter 3, is the full König's Lemma, which states that every infinite, finitely branching tree has an infinite path.) Thus we would say for instance that the theorem that every field has an algebraic closure is computability theoretically weaker than WKL, while the theorem that every commutative ring has a maximal ideal is stronger than WKL (and hence than the theorem that every commutative ring has a prime ideal). As we will see in this book, there are many ways in which computability theoretic notions can be used to make such comparisons between theorems.

We can also use computability theory to make a direct comparison between two principles of the form we have been discussing. Let $P$ and $Q$ be two such principles. Suppose that we can show that, from any instance $X$ of $P$, we can computably obtain an instance $\widehat{X}$ of $Q$ such that, from any solution to $\widehat{X}$, we can computably obtain a solution to $X$. Then we can say we have reduced $P$ to $Q$, and that, in the computability theoretic context, $Q$ implies $P$.

We will be more precise about this and related notions in Section 2.2, but for now, let us give an example. Consider Weak König's Lemma and Lindenbaum's Lemma, which states that every consistent set of sentences (in a given first order language) can be extended to a complete consistent theory. Suppose that we are given such a set of sentences $\Gamma$. Using $\Gamma$, we can effectively enumerate the set $P$ of all sentences provable from $\Gamma$. Now let $\theta_0, \theta_1, \ldots$ be a listing of all sentences in our language. For a sentence $\theta$, let $\theta^0 = \neg\theta$ and $\theta^1 = \theta$. For a binary string $\sigma$, let $\theta_\sigma = \bigwedge_{i<|\sigma|} \theta_i^{\sigma(i)}$. Let $T$ be the tree consisting of all binary strings $\sigma$ such that there is no initial segment $\tau$ of $\sigma$ with $\neg\theta_\tau$ among the first $|\sigma|$ many elements enumerated into $P$. It is not difficult to see that we can obtain $T$ effectively from $\Gamma$, that the consistency of $\Gamma$ implies that $T$ is infinite, and that if $\alpha$ is an infinite path on $T$, then $\{\theta_i^{\alpha(i)} : i \in \mathbb{N}\}$, which can be obtained effectively from $\alpha$, is a completion of $\Gamma$. Thus we say that Weak König's Lemma implies Lindenbaum's Lemma.

In this case, we can also go the other way: Given an infinite binary tree $T$, working in a language with unary relation symbols $R_0, R_1, \ldots$ and a constant symbol $c$, let $\Gamma$ be the set of all sentences of the form $\neg \bigwedge_{i<|\sigma|} R_i^{\sigma(i)}(c)$ for $\sigma \notin T$ (where the superscript notation is as before). Let $C$ be a completion of $\Gamma$ and define $\alpha$ by $\alpha(i) = j$ iff $R_i^j(c) \in C$. Then $\alpha$ is an infinite path on $T$.

Thus WKL and Lindenbaum's Lemma are in fact computability theoretically equivalent, which allows us to say that (at least up to computable operations) WKL represents the combinatorial core of Lindenbaum's Lemma. This equivalence is nontrivial in the sense that these principles are not computably true: as mentioned above, there is a computable infinite binary tree with no computable infinite path, or, equivalently, there is a computable consistent set of formulas with no computable completion (see Corollary 3.7 below).

It is also reasonable to consider the possibility that we might be able to solve $P$ with several applications of $Q$, rather than just one. To formalize this notion, we can consider contexts that are computability theoretically closed, in the sense that if we have access to an object or finite collection of objects $X$, then we have access to any object that is computable from $X$. A Turing ideal is a collection of sets of natural numbers with this property. (We will give a formal definition in Section 2.2.) Say that $P$ holds in a Turing ideal $I$ if for every instance $X \in I$ of $P$, there is a solution $Y \in I$ to $X$. Suppose that for every Turing ideal $I$, if $Q$ holds in $I$ then $P$ holds in $I$. Then it still makes sense to say that $Q$ implies $P$ computability theoretically, albeit in a more general sense than that of our previous notion.

The collection of all computable sets is a Turing ideal, so if $Q$ is computably true and $P$ has a computable instance without computable solutions, then $Q$ does not imply $P$ in the above sense. Thus, for example, the existence of algebraic closures for fields does not imply WKL. We will see in Section 4.5 that there are Turing ideals in which WKL holds that do not contain the halting problem, which shows for instance that WKL does not imply the existence of maximal ideals for commutative rings, even in our more general sense. But at this point we have come quite close to the viewpoint of reverse mathematics, so before considering any more examples, let us discuss that program.

It should also be noted that there are many other topics that fall under the rubric of computable mathematics. For example, there are several lines of research concerned with understanding the relationships between computable copies of a given structure that, while classically isomorphic, are not computably isomorphic. For instance, the *computable dimension* of a structure $\mathcal{M}$ is the number of computable copies of $\mathcal{M}$ up to computable isomorphism. (Without getting into precise definitions, and working in a finite language, a computable structure is one in which the domain and the relevant functions and relations are all computable. A computable ring, for example, is one where the domain is computable, and so are the addition and multiplication operations.) There is a wealth of results on the computable dimension of various structures. As a simple example, for an algebraically closed field $F$, if the transcendence degree of $F$ is finite, then it has computable dimension 1, while if the transcendence degree of $F$ is infinite, then it has computable dimension $\omega$ (see [136,157]). We will restrict ourselves to looking at the kinds of results in computable mathematics that fit the theme of this book, and in particular are connected with reverse mathematics. A much broader picture of the field can be found in [55].

## 1.3 Reverse mathematics

Another approach to calibrating the strength of mathematical principles is to work over a weak base theory. One context in which most mathematicians are familiar with this practice is that of consequences of the Axiom of Choice (AC). It is well-known that Zorn's Lemma, for example, is not just provable from AC, but equivalent to it (as are many other mathematical principles, such as Tychonoff's Theorem and the fact that every vector space (of any cardinality) has a basis). When we say that a mathematical statement is true, we typically mean that it can be proved using the tools generally accepted by the mathematical community, which is generally understood to include AC. But, of course, if we assume AC it is not very meaningful to assert that Zorn's Lemma implies AC. We all understand, however, that what is meant by that statement is that we can prove AC using Zorn's Lemma without appealing to AC itself, or any of its other equivalents. If forced to be precise about it, we might say that the statement that Zorn's Lemma implies AC is provable in ZF, i.e., the usual system ZFC of formal set theory with AC removed. While this level of precision may not be necessary in this case, it is more important when establishing negative results, such as the fact that the statement that every field (of any cardinality) has an algebraic closure, while provable in ZFC, is neither provable in ZF nor implies AC over ZF. (Howard and Rubin [94] lists a large number of consequences of AC and the known implications and nonimplications between them.) In this context, we think of ZF as a "weak theory", i.e., a subsystem of the one in which we ordinarily work, which by virtue of its weakness can be used to prove implications and nonimplications between principles that are all provable (and hence trivially equivalent) given the full power of our accepted methods of proof.

We can see this practice as a form of reverse mathematics: The logical axiom AC can be used to prove theorems in combinatorics, topology, algebra, and so on. Working over a base theory, we can also prove AC from some of these theorems, which shows that the use of choice in their proofs is not just a convenience, but essential. Indeed, we do not tend to draw a real distinction between, say, AC and Zorn's Lemma. For other theorems, we might be able to prove that AC is in fact not essential in proving them, though the base theory itself is not enough. In some cases we might find a weaker logical axiom, such as Dependent Choice, that can be proved equivalent to our given theorem. Proving mathematical theorems from logical axioms is standard practice. Proving logical axioms from mathematical theorems is reverse mathematics. (Though we will take a somewhat broader view here, considering reverse mathematics to be a general practice of proving implications and nonimplications between theorems, be they logical ones or ones arising from other areas of mathematics. As mentioned above, our focus will be particularly on combinatorial principles.)

While ZF may serve as a useful base theory in some cases, it is still rather strong, and clearly not well suited to analyzing the relative strength of principles that use much less of the full machinery of set theory than the ones mentioned above (i.e., principles living in Simpson's realm of "non-set-theoretic mathematics" mentioned in Section 1.1). Setting up an appropriate environment for reverse mathematics at this weaker level involves choosing three things: a language, a logic, and a base theory.

We will work in the language of second order arithmetic, which is actually a two-sorted first order language, with one sort of variables intended to range over natural numbers and another intended to range over sets of natural numbers, the usual symbols of first order arithmetic, and a symbol for set membership. (See Chapter 4 for details.) As mentioned in the previous section, we can use natural numbers and sets of natural numbers to code other mathematical objects, and hence develop a great deal of mathematics in second order arithmetic (always excepting essentially uncountable parts of mathematics). We will discuss this coding process to some extent in Section 4.1, but Simpson [191] describes it in much greater detail. One advantage of working in this setting is that structures in this language are easy to work with, and in particular can often be constructed and studied with computability theoretic methods. We will say more on this topic shortly.

As is almost always done in mathematical practice, we will use classical logic, although computable mathematics is of course related to constructive mathematics. See Section 4 of Bridges and Palmgren [12] for a discussion of the field of constructive reverse mathematics.

As to the choice of base theory $T$, there are a few natural desiderata. We wish to use $T$ to prove theorems of the forms $T \vdash Q \rightarrow P$ and $T \vdash \neg(Q \rightarrow P)$ for (formal versions of) mathematical principles or collections of principles $P$ and $Q$. The weaker $T$ is, the stronger our implication results become, and the finer the distinctions our nonimplication results allow us to make. On the other hand, if $T$ is too weak, we may not be able to prove any nontrivial implications, and our nonimplications may start to become too dependent on extraneous details. The issue of coding provides us with good examples of what would count as "extraneous details". We have mentioned that one way to code a sequence of natural numbers $(n_0, \ldots, n_{k-1})$ is as $\prod_{i<k} p_i^{n_i+1}$. If we are considering only binary sequences, another reasonable way to code $(n_0, \ldots, n_{k-1})$ is as the number whose binary representation is $1n_0 \ldots n_{k-1}$. Suppose $P$ is a principle involving binary sequences. Say we express $P$ as a sentence $\Phi$ in the language of second order arithmetic using our first coding, and as another sentence $\Psi$ using our second coding. We would not want to work over a base theory $T$ in which we could not show that $T \vdash \Phi \leftrightarrow \Psi$. We would also want $T$ to be able to prove basic properties of our codings without which we can do very little (e.g., that there is a function taking (the code of) a finite sequence to its length).

So we need a base theory $T$ that is weak but not too weak. It should also be tractable. Ideally, we should be able to prove theorems of the form $T \vdash Q \rightarrow P$ without having to write down formal proofs. Theorems of the form $T \vdash \neg(Q \rightarrow P)$ are usually proved by exhibiting a model of $T + Q$ that is not a model of $P$, so models of $T$ should be relatively easy to understand and construct. Finally, we would want $T$ to be "natural". From a foundational point of view, we would like provability over $T$ to have some philosophical meaning. From a combinatorial one, when we say that $P$ and $Q$ are equivalent over $T$, we are saying that $P$ and $Q$ have the same "fundamental combinatorics" up to the combinatorial procedures that can be performed in $T$, so we would like this class of procedures to be one we can understand and think of as natural in some sense.

The usual choice of base theory for reverse mathematics is called $\mathrm{RCA}_0$. It consists of first order axioms stating the basic properties of addition, multiplication, and order on the natural numbers; a limited amount of induction; and comprehension (i.e., set existence) axioms just strong enough to imply the existence of all computable sets, or, more precisely, the fact that if sets $X_0, \ldots, X_n$ exist, then so does any set computable from them. (We will give a precise definition of $\mathrm{RCA}_0$ in Section 4.1.) This system has proved to fit our criteria quite well. It is weak enough to make many fine distinctions like the ones mentioned in the computability theoretic context in the previous section, but strong enough to avoid making meaningless ones. As we will see, it is not difficult (with a bit of practice) to argue informally about what is provable in $\mathrm{RCA}_0$. The naturality of $\mathrm{RCA}_0$ is argued for by the fundamental nature of the notion of computability, as well as the close connections between computability and definability, as expressed for instance in Post's Theorem 2.3 below.

Finally, a structure $\mathcal{M}$ in the language of second order arithmetic consists of a structure $M$ in the language of first order arithmetic together with a second order part consisting of a collection of subsets of the domain of $M$. We will see that when $M$ is just the usual structure of the natural numbers, $\mathcal{M}$ is a model of $\mathrm{RCA}_0$ iff it is a Turing ideal, as defined in the previous section. Thus the computability theoretic approach described above is very close to the reverse mathematical one, and the large collection of tools developed for computability theoretic analysis can be brought to bear on constructing and studying models of $\mathrm{RCA}_0$ (even for models with nonstandard first order parts, since we can generalize computability theoretic methods to those cases with some care).

Having taken a brief look at the computability theoretic and reverse mathematical points of view, and at how they are connected, let us now proceed to see how they are employed in practice.

## 1.4   An overview

After reviewing some of the basics of computability theory and introducing the idea of forcing in Chapter 2, in Chapter 3 we examine the strength of versions of König's Lemma, as an example of the computability theoretic approach. We then give an introduction to reverse mathematics in Chapter 4. Chapter 5 is a discussion of the nature of the subtle structure I see in the kind of work in computable mathematics and reverse mathematics that I do.

Chapter 6 is in a sense the heart of this book. In it, we explore the computability theory and reverse mathematics of versions of Ramsey's Theorem, which constitute our central case study. Here Ramsey's Theorem is the statement that for any $n \geqslant 1$ and any coloring of the $n$-element subsets of $\mathbb{N}$ with finitely many colors, there is an infinite set $H$ such that all $n$-element subsets of $H$ have the same color. In Chapter 7, we study the powerful method of conservativity results, continuing our case studies of versions of König's Lemma and of Ramsey's Theorem. In Chapter 8, we summarize many of the results discussed in previous chapters as diagrams.

Ramsey's Theorem for Pairs (the $n = 2$ case above) is particularly interesting, and there is a whole universe of principles that follow from it, some of which we look at it in Chapter 9. These principles include some theorems of basic model theory, such as the Atomic Model Theorem (AMT), which states that every complete atomic theory has an atomic model. Discussing such principles might seem somewhat off-topic, but model theoretic principles such as AMT fit quite well into our universe of combinatorial principle.

Indeed, much like Lindenbaum's Lemma, whose combinatorial character is revealed by its equivalence with WKL, AMT can be reinterpreted (in the precise sense of equivalence over $RCA_0$) as a theorem about paths on trees. Finally, in Chapter 10, we discuss several topics that complement the ones in the rest of the book and point to the continued richness of this area of research.

I have tried to give attributions and references for the theorems and exercises below, but in some cases these results are folklore, or have become well-known enough to be generally quoted without citations, and I have not been able to trace down the original sources. In any case, none of the results below are original to this book.

## 1.5 Further reading

There are many expository / survey papers and books related to the topic of this book. The following list is by no means exhaustive, but should give those looking to learn more about the subject a good start.

The two volumes of the *Handbook of Recursive Mathematics* [55] cover a wide range of topics in computable mathematics. We will have occasion to cite in particular the articles by Cenzer and Remmel [15] on $\Pi_1^0$ classes, Downey [39] on computable linear orders, and Harizanov [77] on computable model theory. Other articles in those volumes related to our topics include the ones by Gasarch [70] on computable combinatorics and Simpson and Rao [193] on the reverse mathematics of algebra.

The standard textbook in reverse mathematics is Simpson's classic *Subsystems of Second Order Arithmetic*, now in its second edition [191], which will be referred to several times below. It is definitely the place to go for a thorough grounding in the area. Simpson has also edited a collection, *Reverse Mathematics 2001* [189], which gives a picture of the diversity of the field. At the time of writing, Dzhafarov and Mummert are working on a book, to be published by Springer in the Theory and Applications of Computability series, which should be an excellent complement to Simpson's book.

As mentioned above, two articles by Simpson that discuss the foundational import of reverse mathematics (as does his book) are [187] and [190]. There is of course a great deal to say about the foundations of mathematics and what metamathematical programs such as reverse mathematics have to do with them, but these articles can serve as a good starting point for those interested in these issues.

Shore's "Reverse mathematics: the playground of logic" [183], an extended version of his Gödel Lecture at the 2009 Logic Colloquium in Sofia, is an excellent invitation to the field from a point of view similar to that of this book. It also discusses one of the current frontiers of the field, extending reverse mathematics to the uncountable setting, as does his paper [184]. Other papers in *Effective Mathematics of the Uncountable* [73], edited by Greenberg, Hamkins, Hirschfeldt, and Miller, discuss various approaches to extending computable mathematics to the uncountable setting.

Kohlenbach [112] proposes a higher order reverse mathematics, i.e., one that explicitly deals with third and higher order objects.

In December, 2008, a workshop on Computability, Reverse Mathematics, and Combinatorics was held at the Banff International Research Station, resulting in a list of open problems [1]. Montalbán [146] also discusses open problems in several areas of reverse mathematics.

Reading the original papers discussed in expository works such as this one is of course also important. Jockusch [97] and Cholak, Jockusch, and Slaman [20], for example, are classics that will amply reward the reader. Other major papers in this line of research, including more recent ones answering open questions in [20], will be mentioned below.

## Chapter 2

# Gathering Our Tools: Basic Concepts and Notation

In this chapter, we first review a few essential computability theoretic concepts, as a reminder and to fix notation; for more information, see [158, 159, 170, 196, 197] or Chapter 2 of [40]. This review assumes knowledge of basic concepts such as computable (also known as recursive) sets and functions, computably enumerable (also known as recursively enumerable) sets, and Turing reductions. We then introduce the important technique of forcing, originally developed in the context of set theory but of great usefulness in computability theory and reverse mathematics.

## 2.1 Computability theory

Unless otherwise specified or clear from the context, when we say "set" and "function" we mean a set of natural numbers and a function $\mathbb{N} \to \mathbb{N}$, respectively, and we use variables such as $X$ and $f$ for such sets and functions, respectively.

We fix an effective one-to-one listing $D_0, D_1, \ldots$ of the finite sets (of natural numbers), and call $n$ the *canonical index* of $D_n$. ("Effective" here means that we can computably determine $D_n$ given $n$.)

A *partial function* $\varphi : \mathcal{X} \to \mathcal{Y}$ is one whose domain is a subset of $\mathcal{X}$. If the domain is in fact all of $\mathcal{X}$, then $\varphi$ is *total*. For a partial function $\varphi$, we write $\varphi(n)\downarrow$ if $\varphi(n)$ is defined and $\varphi(n)\uparrow$ otherwise.

For a set $X$, we write $X^{<\mathbb{N}}$ for the set of finite sequences of elements of $X$, and $X^{\mathbb{N}}$ for the set of infinite sequences of elements of $X$. (In computability theory, it is more usual to write $X^\omega$ and $X^{<\omega}$. However, in reverse mathematics we sometimes work over nonstandard models of fragments of Peano Arithmetic, as will be discussed in Chapter 4 below, and we want to reserve $\omega$ for the standard natural numbers (cf. Definition 4.18). Of course, when proving purely computability theoretic theorems, we do work in the standard structure, but we want our notation to be flexible enough to be employed in the reverse mathematical context as well. In general, we will use $\omega$ for the natural numbers only when specifically emphasizing the distinction between the standard natural numbers and possibly non-standard structures.) For $\alpha \in X^{\mathbb{N}}$, we write $\alpha(n)$ for the $(n+1)$st element of the sequence $\alpha$, and $\alpha \upharpoonright n$ for the string $\alpha(0) \ldots \alpha(n-1) \in X^{<\mathbb{N}}$. We identify sets of natural numbers with elements of $2^{\mathbb{N}}$, and thus think of elements of $2^{<\mathbb{N}}$ as initial segments of such sets.

We write $X \leqslant_{\mathrm{T}} Y$ to mean that $X$ is computable in $Y$, in the sense already mentioned in Section 1.2. As mentioned in that section, we also say that $X$ is computable relative to $Y$ or $Y$-computable. In addition, we say that $X$ is *Turing reducible* to $Y$. This notion of reducibility gives rise to an equivalence relation $\equiv_{\mathrm{T}}$, whose equivalence classes are the *(Turing) degrees*. Two sets in the same Turing degree are said to be *Turing equivalent*. (We define Turing reducibility and Turing equivalence for functions or other countable objects similarly.) For $A, B \subseteq \mathbb{N}$, let $A \oplus B = \{2n : n \in A\} \cup \{2n + 1 : n \in B\}$. The degree of $A \oplus B$ is the least upper bound of the degrees of $A$ and $B$.

Using a formalism such as Turing machines with oracle tapes, we may similarly define the notion of $X$ being *c.e. relative to* $Y$, or *c.e. in* $Y$, or *$Y$-c.e.* (We write "c.e." as an abbreviation for "computably enumerable".)

We fix an effective list $\Phi_0, \Phi_1, \ldots$ of the Turing functionals (i.e., Turing reduction procedures). Then $\Phi_0^X, \Phi_1^X, \ldots$ is an effective list of all partial $X$-computable functions. ("Effective" here means that there is a partial $X$-computable function $U$ such that $U(e, n) = \Phi_e^X(n)$ for all $n$.) We let $W_e^X = \mathrm{rng}\, \Phi_e^X$, so that $W_0^X, W_1^X, \ldots$ is an effective list of all $X$-c.e. sets. We write $\Phi_e$ and $W_e$ for $\Phi_e^{\emptyset}$ and $W_e^{\emptyset}$, respectively. We identify a set of natural numbers with its characteristic function, so we think of a total $0, 1$-valued $\Phi_e$ as a computable set. The *use* of a convergent oracle computation is the least $u$ such that the computation queries its oracle only on numbers less than $u$. The *use principle* is the simple but important fact that if $\Phi^X(n)\downarrow$ with use $u$ and $Y \upharpoonright u = X \upharpoonright u$, then $\Phi^Y(n)\downarrow = \Phi^X(n)$.

Let $X$ be c.e. We write $X[s]$ for the set of numbers enumerated into $X$ by stage $s$. For a Turing functional $\Phi$, we write $\Phi^Z[s]$ for the result of carrying out $s$ many steps of the computation of $\Phi$ with oracle $Z$.

The following theorem, known as the Recursion Theorem, is one of the fundamental facts of computability theory. It allows us to use an index for a computable function as part of the definition of that function. It thus forms the theoretical underpinning of the common programming practice of having a routine make recursive calls to itself. See e.g. Soare [196] for a proof.

**Theorem 2.1** (Kleene [110]). *Let $f$ be a total computable function. Then there is an $e$ such that $\Phi_e = \Phi_{f(e)}$.*

The *halting problem relative to* $X$ is $X' = \{e : \Phi_e^X(e)\downarrow\}$. Thus the unrelativized halting problem is denoted by $\emptyset'$. We also refer to $X'$ as the *jump* of $X$. We use the following notation for iterates of the jump: $X^{(0)} = X$ and $X^{(n+1)} = (X^{(n)})'$. We usually write $X''$ for $X^{(2)}$. An $\emptyset'$-computable set is *complete* if it is Turing equivalent to the halting problem, and *incomplete* otherwise.

A set $X$ $low_n$ if $X^{(n)} \equiv_T \emptyset^{(n)}$ and $high_n$ if $X^{(n)} \geqslant_T \emptyset^{(n+1)}$. If $n = 1$, we write simply "low" and "high", respectively. A *lowness index* for a low set $A$ is an $e$ such that $\Phi_e^{\emptyset'} = A'$. An important fact about the double jump is that $\emptyset''$ can decide whether a given computable set is infinite. The following fact will be useful below; proofs can be found in [40, 196]. A function $f$ *dominates* a function $g$ if $f(n) \geqslant g(n)$ for all but finitely many $n$.

**Theorem 2.2** (Martin [128]). *A set $X$ is high iff it computes a function that dominates all (total) computable functions. (Such a function is called a* dominant *function.)*

We define the *arithmetic hierarchy* as follows. A set $A$ is $\Sigma_n^0$ if there is a computable relation $R(x_0, \ldots, x_{n-1}, y) \subseteq \mathbb{N}^{n+1}$ such that $y \in A$ iff

$$\exists x_0 \, \forall x_1 \, \exists x_2 \, \forall x_3 \cdots Q x_{n-1} \, R(x_0, \ldots, x_{n-1}, y). \tag{2.1}$$

Since the quantifiers alternate, $Q$ is $\exists$ if $n$ is odd and $\forall$ if $n$ is even. In this definition, we could have had $n$ alternating quantifier blocks, instead of single quantifiers, but we can always collapse two successive existential or universal quantifiers into a single one by using pairing functions, so that would not make a difference. The definition of $A$ being $\Pi_n^0$ is the same, except that the leading quantifier is a $\forall$ (but there still are $n$ alternating quantifiers in total). It is easy to see that $A$ is $\Pi_n^0$ iff its complement $\overline{A}$ is $\Sigma_n^0$. Finally, we say a set is $\Delta_n^0$ if it is both $\Sigma_n^0$ and $\Pi_n^0$ (or equivalently, if both it and its complement are $\Sigma_n^0$).

Note that the $\Delta_0^0$, $\Pi_0^0$, and $\Sigma_0^0$ sets are all exactly the computable sets. The same is true of the $\Delta_1^0$ sets, by the $n = 0$ case of the following theorem, which is a key fact in connecting computability theoretic concepts with ones based on definability, and is known as Post's Theorem. (For a proof see for instance [40, 196].)

**Theorem 2.3** (Post [165]). *A set is $\Sigma^0_{n+1}$ iff it is c.e. in $\emptyset^{(n)}$, and is $\Delta^0_{n+1}$ iff it is computable in $\emptyset^{(n)}$.*

Clearly, every $\Sigma^0_n$ or $\Pi^0_n$ set is $\Delta^0_{n+1}$. For $n > 0$, the fact that this containment is proper follows from the above theorem and the fact that, for any $X$, there are sets computable in $X'$ that are neither $X$-c.e. nor $X$-co-c.e., for instance $X' \oplus \overline{X'}$. Similarly, the fact that there are $\Sigma^0_n$ sets that are not $\Pi^0_n$, and vice-versa, comes from the fact that, for any $X$, there are $X$-c.e. sets that are not $X$-computable (together with the fact that if a set and its complement are both $X$-c.e., then they are $X$-computable).

The $\Delta^0_2$ sets can also be characterized as the ones that are computably approximable. This fact is known as the limit lemma. (For a proof see for instance [40, 196].)

**Theorem 2.4** (Shoenfield [180]). *A set $A$ is $\Delta^0_2$ iff there is a computable $0, 1$-valued binary function $g$ such that $\lim_s g(n, s)$ exists and $A(n) = \lim_s g(n, s)$ for all $n$.*

When we are given a $\Delta^0_2$ set $A$, we assume we have fixed a $g$ as in the limit lemma and write $A(n)[s]$ for $g(n, s)$. We think of $A[s] = \{n : A(n)[s] = 1\}$ as the stage $s$ approximation to $A$. There is also the following strong form of the limit lemma for higher levels of the arithmetic hierarchy. It is a straightforward but useful exercise to prove it using Post's Theorem and the relativized form of the limit lemma (see below for a discussion of relativization).

**Theorem 2.5** (Shoenfield [180]). *Let $k \geqslant 2$. A set $A$ is $\Delta^0_k$ iff there is a computable $0, 1$-valued $k$-ary function $g$ such that for all $n$,*
$$A(n) = \lim_{s_1} \lim_{s_2} \ldots \lim_{s_{k-1}} g(n, s_1, s_2, \ldots, s_{k-1}).$$

We can also define the *analytic hierarchy* of $\Sigma^1_n$, $\Pi^1_n$, and $\Delta^1_n$ sets, but we will not need it here. In a few places, we will mention the *hyperarithmetic* sets, which coincide with the $\Delta^1_1$ sets. Since we will not need any of the theory of hyperarithmetic sets, we will not discuss them further here. Definitions and basic facts can be found for instance in Sacks [175].

We have already encountered the important idea of *relativization* in concepts such as computability relative to a set and the relativized halting problem. This idea will be used repeatedly below. To relativize a theorem of computability theory to a set $X$ is to replace "computable" by "$X$-computable", "c.e." by "$X$-c.e.", and in general, any given computability theoretic concept by the analogous concept relative to the oracle

$X$. Most computability theoretic results remain true in relativized form, with essentially the same proofs. For example, we can define the levels of the arithmetic hierarchy relative to a set $X$ as above, but with an $X$-computable relation $R$ in (2.1). Then a set is $\Sigma^0_{n+1}$ relative to $X$ iff it is $X^{(n)}$-c.e., and similarly for the other results mentioned above. Following standard computability theoretic practice, we will often state theorems in unrelativized form and later use their relativized forms. (For a simple example, see Proposition 2.7 below, and the comment following its proof.) This practice helps simplify the statements and proofs of theorems, and is justified by the fact that, in these cases, stating the relativized forms of the theorems and adapting their proofs is straightforward. However, one does sometimes have to be careful to make sure one is relativizing theorems correctly. For example, consider the relativization of Theorem 2.2 to a set $Y$. If $X \geqslant_{\mathrm{T}} Y$, then $X' \geqslant_{\mathrm{T}} Y''$ iff $X$ computes a function that dominates all $Y$-computable functions. But the relativized form of that theorem for a general $X$ is that $(X \oplus Y)' \geqslant_{\mathrm{T}} Y''$ iff $X \oplus Y$ computes a function that dominates all $Y$-computable functions. When encountering uses of relativization in the text below, readers unfamiliar with this process should write out the relativized forms of the relevant theorems and proofs in a few cases as an exercise.

## 2.2    Computability theoretic reductions

Let us now give formal definitions of the notions of computability theoretic reduction discussed in Section 1.2. By a *problem* we mean a true principle $P$ of the form $\forall X \, [\theta(X) \rightarrow \exists Y \, \psi(X,Y)]$, where $\theta$ and $\psi$ are arithmetic, i.e., they do not involve quantification over second order objects. (See Chapter 4 for a formal description of the language of second order arithmetic.) As mentioned in Section 1.2, an *instance* of $P$ is an $X$ such that $\theta(X)$ holds, and a *solution* to this instance is a $Y$ such that $\psi(X,Y)$ holds. For the remainder of this section, let $P$ and $Q$ be problems.

It is worth noting that an informally stated problem might have more than one reasonable formalization. Consider Weak König's Lemma (WKL). As mentioned in Section 1.2, it is natural to say that an instance of WKL is an infinite binary tree $T$, and a solution to $T$ is an infinite path on $T$. However, this definition depends on our exact choice of coding of trees and paths as sets of natural numbers. Furthermore, we could also consider every binary tree $T$ to be an instance of WKL, and say that, if $T$ is infinite, then a solution to $T$ is an infinite path on $T$, while if $T$ is finite then any set counts as a solution to $T$. In some situations, the exact choice of formalization might matter, and must then be made explicit. In this book, however, it should always be clear what the most natural formalization of any given informally stated problem is, and hence what its instances and solutions are, at least up to the choice of codings for the first and second order objects mentioned in the principle. Furthermore, such choices of codings will not make a difference to the results we discuss.

We say that $P$ is *computably reducible* to $Q$, and write $P \leqslant_c Q$, if for any instance $X$ of $P$, there is an $X$-computable instance $\widehat{X}$ of $Q$ such that, for any solution $\widehat{Y}$ to $\widehat{X}$, there is an $(X \oplus \widehat{Y})$-computable solution $Y$ to $X$.

The reason for having $Y$ be $(X \oplus \widehat{Y})$-computable, rather than just $\widehat{Y}$-computable, might not be immediately clear, but consider the following trivial example. Let $P$ be $\forall X \, \exists Y \, Y = X$, the "problem" of obtaining $X$ given $X$. Clearly we want to say that $P$ is implied by every true principle $Q$, since we can solve $P$ without even looking at $Q$. Now let $Q$ be $\forall X \, \exists Y \, Y = \emptyset$, the equally trivial problem of obtaining $\emptyset$ given $X$. If $X$ is not computable, then we cannot obtain a solution to the instance $X$ of $P$ effectively from a solution to any instance $\widehat{X}$ of $Q$ (i.e., we cannot obtain a noncomputable set effectively from $\emptyset$), but we can obtain one effectively from such a solution (even for $\widehat{X} = \emptyset$) together with $X$. Nevertheless, the notion of reducibility between principles where we do not use the power of the instance $X$, which is known as *strong computable reducibility*, is still of interest in some cases; see Dzhafarov [47] and Hirschfeldt and Jockusch [85]. (Of course, in many cases, the two notions coincide, for example when we can code $X$ into $\widehat{X}$ so that any solution to $\widehat{X}$ computes $X$.)

Also of interest is the idea of reducing $P$ uniformly to $Q$. We say that $P$ is *uniformly reducible* to $Q$, and write $P \leqslant_u Q$, if there are Turing functionals $\Phi$ and $\Psi$ such that if $X$ is an instance of $P$, then $\widehat{X} = \Phi^X$ is an instance of $Q$, and if $\widehat{Y}$ is a solution to $\widehat{X}$, then $Y = \Psi^{X \oplus \widehat{Y}}$ is a solution to $X$. We will see examples below where $P \leqslant_c Q$ but $P \not\leqslant_u Q$. Uniform reducibility is equivalent to a special case of the notion of Weihrauch reducibility, introduced by Weihrauch [208, 209] in the context of computable analysis, and widely studied since, including for purposes of computability theoretic comparison of mathematical principles; see for instance Brattka and Gherardi [10, 11]. Therefore, it is sometimes denoted by $\leqslant_W$ instead of $\leqslant_u$. For further discussion of Weihrauch reducibility in our context, and a proof of equivalence between uniform reducibility and (a special case of) Weihrauch reducibility, see Dorais, Dzhafarov, Hirst,

Mileti, and Shafer [38]. Again, we can also consider a notion of *strong uniform reducibility* (also known as strong Weihrauch reducibility), where the above definition is changed by having $Y = \Psi^{\widehat{Y}}$; see for instance [38,85].

Note that, although looking at all instances of a problem, rather than just computable ones, in the above definitions seems better suited to capture the idea of reducing a problem to another, the usual way to show that $Q$ does not imply $P$ is to exhibit a computable instance $X$ of $P$ with no corresponding computable $\widehat{X}$ as above.

As discussed in Section 1.2, we also want to consider a more general notion of reducibility defined using Turing ideals. A nonempty $I \subseteq \mathcal{P}(\mathbb{N})$ is a *Turing ideal* if the following hold for all $X$ and $Y$.

(i) If $X, Y \in I$ then $X \oplus Y \in I$.
(ii) If $X \in I$ and $Y \leqslant_T X$ then $Y \in I$.

We say that $P$ *holds* in $I$ if every instance $X \in I$ of $P$ has a solution $Y \in I$. Two notations have been proposed to denote the property that for every Turing ideal $I$, if $Q$ holds in $I$ then so does $P$. Shore [183, 184] writes $Q \vDash_c P$, while Hirschfeldt and Jockusch [85] write $P \leqslant_\omega Q$. As mentioned in Section 1.3 and discussed more formally in Section 4.5 below, Turing ideals are exactly the second order parts of models of the usual weak base system $\mathrm{RCA}_0$ of reverse mathematics with standard first order part. Such a model is called an $\omega$-model, so in the above situation, we will simply say that every $\omega$-model of $Q$ is an $\omega$-model of $P$. For a uniform version of this notion, see Hirschfeldt and Jockusch [85].

## 2.3 Forcing

We will need only the basic apparatus of the theory of forcing, in our particular setting of second order arithmetic. We will not need this material until Section 6.5, and will not need the forcing relation itself until Chapter 7.

Forcing arguments, and their effective analogs discussed below, may be seen as generalizations of finite extension arguments such as the proof by Kleene and Post [111] (see also Theorem VI.1.2 of [196]) that there are incomparable degrees below the degree of $\emptyset'$. A *notion of forcing* is a partial order $(P, \preccurlyeq)$. We call the elements of $P$ *conditions* and say that $p$ *extends* $q$ if $p \preccurlyeq q$. The idea is that conditions represent partial information about some object $G$ we want to build; if $p$ extends $q$ then $p$ contains at least the same information about $G$ as $q$, and possibly more. (The reason for writing $p \preccurlyeq q$ in this case is that we identify $p$ with the collection of objects that $G$ could be, given the information contained in $p$; we say that such objects are *compatible* with $p$. More information means a smaller range of possible objects. Some authors reverse the notation.) For example, in *Cohen forcing* the conditions are finite partial functions from $\mathbb{N}$ to 2, and $p \preccurlyeq q$ if $p \supseteq q$, while the object $G$ is a function from $\mathbb{N}$ to 2. Such a $G$ is compatible with a condition $p$ if $G(n) = p(n)$ for all $n \in \operatorname{dom} p$.

A subset $D \subseteq P$ is *dense* if for every $p \in P$, there is an element of $D$ that extends $p$. We think of a dense set as representing a requirement on the object $G$ we are building (i.e., we require that the object *meet* $D$, i.e., be compatible with at least one element of $D$). Density ensures that, no matter how much partial information we have about $G$, it is still possible for it to meet $D$. For example, in the case of Cohen forcing, for a fixed function $f : \mathbb{N} \to 2$, we can let $D_f = \{p : \exists n\,[p(n) \neq f(n)]\}$. Then $D_f$ is clearly dense, and represents the requirement that $G \neq f$. Any $G$ meeting $D_f$ satisfies this requirement.

A subset $F \subseteq P$ is a *filter* if it is closed upwards (i.e., $p \in F \wedge p \preccurlyeq q \to q \in F$) and any two elements of $F$ have a common extension in $F$ (i.e., $p, q \in F \to \exists r \in F\,[r \preccurlyeq p, q]$). Let $\mathcal{D}$ be a collection of dense subsets of $P$. A filter $F$ is $\mathcal{D}$-*generic* if it meets every element of $\mathcal{D}$, i.e., for each $D \in \mathcal{D}$, we have $D \cap F \neq \emptyset$. We think of $\mathcal{D}$ as a collection of requirements on the object $G$ we are building. An object satisfying these requirements can then be obtained from a $\mathcal{D}$-generic filter. For example, in the context of Cohen forcing, we can take a collection $\mathcal{H}$ of functions from $\mathbb{N}$ to 2 and let $\mathcal{D}$ consist of the dense sets $D_f$ for $f \in \mathcal{H}$, where $D_f$ is as above, together with dense sets $E_n$ of the form $\{p : p(n)\!\downarrow\}$. If $F$ is a $\mathcal{D}$-generic filter, then we can let $G = \bigcup F$. The definition of filter ensures that $G$ is a function (i.e., if $p, q \in F$ are both defined at $n$, then it must be the case that $p(n) = q(n)$, as otherwise $p$ and $q$ could not have a common extension in $F$). Meeting the $E_n$ ensures that $G$ is total, while meeting the $D_f$ ensures that $G \neq f$ for all $f \in \mathcal{H}$.

The main fact about the existence of generic filters is the following:

**Proposition 2.6.** *Let $(P, \preccurlyeq)$ be a notion of forcing, let $\mathcal{D}$ be a countable collection of dense subsets of $P$, and let $p \in P$. Then there is a $\mathcal{D}$-generic filter containing $p$.*

*Proof.* Let $D_0, D_1, \ldots$ be the elements of $\mathcal{D}$. Let $p_{-1} = p$, and for each $n$, let $p_n$ be an element of $D_n$ extending $p_{n-1}$. Let $F = \{q : \exists n \, [p_n \preccurlyeq q]\}$. It is easy to check that $F$ is a $\mathcal{D}$-generic filter containing $p$. $\qquad\square$

We usually identify a filter $F$ with the object $G$ that we define using $F$, and say that $G$ is $\mathcal{D}$-generic when $F$ is.

It is sometimes simpler in stating certain facts not to specify a collection of dense sets ahead of time, but instead to use the phrase "sufficiently generic". When we say that every sufficiently generic $G$ has property $\Phi$, we mean that there is a countable collection of dense sets $\mathcal{D}$ such that every $\mathcal{D}$-generic $G$ has property $\Phi$. Note that this definition implies that if every sufficiently generic $G$ has property $\Phi$ and every sufficiently generic $G$ has property $\Psi$, then every sufficiently generic $G$ has property $\Phi \wedge \Psi$, since the union of two countable collections of dense sets is itself a countable collection of dense sets.

In computability theory, one often considers sets that are generic for classes of dense sets defined using levels of the arithmetic hierarchy. For example, a set $G$ is *n-generic* if for every $\Sigma_n^0$ subset $A$ of $2^{<\mathbb{N}}$, there is an initial segment $\sigma$ of $G$ such that either $\sigma \in A$ or $\tau \notin A$ for all $\tau$ extending $\sigma$. We say that $G$ *meets or avoids* every $\Sigma_n^0$ set of strings. (Note that for any set $A$ of strings, the strings $\sigma$ such that either $\sigma \in A$ or $\tau \notin A$ for all $\tau$ extending $\sigma$ form a dense set.) See Jockusch [100] for more on this topic. The theory of algorithmic randomness may also be seen as arising from effectivizing a notion of forcing (see Section 7.2.5 of Downey and Hirschfeldt [40]). We will not be discussing these notions in this book, however.

Proposition 2.6 has an effective analog. Let $(P, \preccurlyeq)$ be a countable notion of forcing and let $c : \mathbb{N} \to P$ be a surjective partial function. We think of each $i$ such that $c(i) = p$ as an index for $p$. For Cohen forcing, for example, a natural choice of $c$ would be to take the canonical index of $\{\langle n, j \rangle : p(n) = j\}$ to $p$. A dense set $D \subseteq P$ is *effectively dense* (with respect to $c$) if there is a partial computable function $f$ such that for each $i \in \operatorname{dom} c$, we have $c(f(i)) \preccurlyeq c(i)$ and $c(f(i)) \in D$. We usually drop the phrase "with respect to $c$" when $c$ has been fixed. Dense sets $D_0, D_1, \ldots$ are *uniformly effectively dense* if there are uniformly partial computable functions $f_0, f_1, \ldots$ witnessing the effective density of $D_0, D_1, \ldots$, respectively. We say that a countable collection of dense sets $\mathcal{D}$ is uniformly effectively dense if there is an ordering $D_0, D_1, \ldots$ of the elements of $\mathcal{D}$ (possibly with repetitions) that makes them uniformly effectively dense. A set $S \subseteq P$ *generates* a filter $F \subseteq P$ if $F = \{q : \exists p \in S \, [p \preccurlyeq q]\}$.

**Proposition 2.7.** *Let $(P, \preccurlyeq)$ be a countable notion of forcing, let $p \in P$, and let $c : \mathbb{N} \to P$ be a surjective partial function. Let $\mathcal{D}$ be a uniformly effectively dense collection of subsets of $P$. Then there is a computable sequence $i_0, i_1, \ldots$ such that $\{c(i_n) : n \in \mathbb{N}\}$ generates a $\mathcal{D}$-generic filter.*

*Proof.* Let $D_0, D_1, \ldots$ be a listing of the elements of $\mathcal{D}$ that makes them uniformly effectively dense. We proceed as in the proof of Proposition 2.6, but obtain $i_n$ computably from $i_{n-1}$ using the effective density of $D_n$.    $\square$

We typically use this proposition in relativized form. That is, we relativize the notion of uniformly effectively dense in the obvious way, and conclude that if $\mathcal{D}$ is a collection of subsets of $P$ that is uniformly effectively dense relative to a given set $X$, then there is an $X$-computable sequence $i_0, i_1, \ldots$ such that $\{c(i_n) : n \in \mathbb{N}\}$ generates a $\mathcal{D}$-generic filter.

The point of the proposition is that, when the function $c$ gives a natural indexing of the conditions (as in the Cohen forcing example mentioned above), the generic object $G$ we wish to construct can often be obtained computably from the sequence $i_0, i_1, \ldots$. For instance, if $\mathcal{H}$ is a countable collection of uniformly computable functions from $\mathbb{N}$ to 2, then letting $D_f$ and $E_n$ be as above, the collection $\mathcal{D}$ of all $D_f$ for $f \in \mathcal{H}$ and all $E_n$ is uniformly effectively dense (with respect to the natural indexing function $c$ for the Cohen forcing conditions mentioned above). Letting $i_0, i_1, \ldots$ be as in Proposition 2.7, we can define a computable function $G$ as follows. Given $n$, search for a $j$ such that $c(i_j)(n)\!\downarrow$ and define $G(n) = c(i_j)(n)$. It is easy to check that $G$ is a total computable function from $\mathbb{N}$ to 2 that is not in $\mathcal{H}$. Thinking of the functions in $\mathcal{H}$ and of $G$ as characteristic functions of subsets of $\mathbb{N}$, we have a (somewhat roundabout) proof that there is no uniformly computable listing of all computable sets.

The following exercise introduces the important notion of *forcing the jump*. For a Cohen forcing condition $p$, we write $\Phi_e^p(e)\!\downarrow$ iff the functional $\Phi_e$ converges in $\leqslant |\mathrm{dom}\, p|$ many stages on input $e$ when using $p$ as an oracle. (If the functional attempts to query a value on which $p$ is undefined, then $\Phi_e^p(e)\!\uparrow$.) Note that, by the use principle, if $\Phi_e^p(e)\!\downarrow$ then $\Phi_e^G(e)\!\downarrow$ for all $G \supset p$.

$\boxed{1}$ **Exercise 2.8.** Let $(P, \preccurlyeq)$ be the notion of Cohen forcing, and let $c$ be the natural indexing function given above. Let $\mathcal{H}$ be the collection of all total computable functions from $\mathbb{N}$ to 2, and let $D_f$ and $E_n$ be as above. Let $J_e$ be the set of all conditions $p$ such that either $\Phi_e^p(e)\!\downarrow$ or $\Phi_e^q(e)\!\uparrow$ for every $q \preccurlyeq p$. Let $\mathcal{D}$ be the collection of all $D_f$ for $f \in \mathcal{H}$, all $E_n$, and all $J_e$.

**a.** Show that $\mathcal{D}$ is uniformly effectively dense relative to $\emptyset'$.

**b.** Let $i_0, i_1, \ldots$ be as in Proposition 2.7 (relativized to $\emptyset'$), and define $G$ as in the second paragraph following the proof of Proposition 2.7. Show that $G$ is a low noncomputable set. [Hint: For lowness, show that $\emptyset'$ can be used to find a $p \subset G$ such that $p \in J_e$, and hence to determine whether $\Phi_e^G(e)\!\downarrow$, using the fact that this is the case iff there is a $q \subset G$ such that $\Phi_e^q(e)\!\downarrow$.]

In the above exercise, what allows us to conclude that $G$ is low is that the question of whether $e \in G'$, which, even given the power of $\emptyset'$ to answer existential questions, in principle requires knowledge of all of $G$, can be reduced to a question about finite conditions that $\emptyset'$ can answer. It is this idea of having compatibility with a finite condition force a given $e$ to be in or out of the jump of $G$ that we mean by the phrase "forcing the jump". For another example of a forcing argument, using a different notion of forcing, see Exercise 3.11 below.

In general, the word "forcing" comes from the situation where, for some property $R$, every sufficiently generic filter containing $p$ satisfies $R$. In this case we say that $p$ *forces* $R$ and write $p \Vdash R$. In simple cases, it might be that *every* filter containing $p$ satisfies $R$. For example, in the context of Cohen forcing, if $\Phi_e^p(e)\!\downarrow$ then $\Phi_e^G(e)\!\downarrow$ for all $G$ compatible with $p$. For more complicated properties, however, we do need to restrict ourselves to sufficiently generic filters. Consider Cohen forcing and let $R$ be the property that $G$ (thought of as a set) is infinite (recall that $G = \bigcup F$ for our generic filter $F$, so this is indeed a property of $F$). It is of course never the case that *every* $G$ corresponding to a filter containing $p$ is infinite, but for every sufficiently generic $F$, the set $G$ is indeed infinite, so we do want to say that $p \Vdash R$ for every $p$ (even $p = \emptyset$). More generally, suppose that for every $q \preccurlyeq p$ and every $m$, there is an $r \preccurlyeq q$ and an $n$ such that $r \Vdash \Phi(G, m, n)$, where $\Phi$ is some given property. If $G$ corresponds to a sufficiently generic filter $F$ containing $p$, then for each $m$, the filter $F$ will contain some $r$ such that $r \Vdash \Phi(G, m, n)$ for some $n$, and hence $\forall m \, \exists n \, \Phi(G, m, n)$ will hold. Thus we have $p \Vdash \forall m \, \exists n \, \Phi(G, m, n)$.

This semantic definition of forcing corresponds to a syntactic one. We will use $\Vdash^*$ for this notion initially, but will drop this notation once we establish that $\Vdash$ and $\Vdash^*$ coincide. In this book, $G$ will always be a set of (possibly nonstandard) natural numbers, and all of the properties we will consider will be expressible in the language of first order arithmetic, together with the symbol $\in$, a set parameter $C$, and the set variable $G$. The

atomic formulas are the atomic formulas of first order arithmetic together
with $t \in C \oplus G$ for a term $t$ (we will also want to consider formulas such as
$t \in G$, but we can think of such a formula as abbreviating $2t + 1 \in C \oplus G$).
Of course, for an atomic sentence of first order arithmetic $\sigma$, we have $p \Vdash^* \sigma$
iff $\sigma$ is true. In all our examples, the meaning of $p \Vdash^* t \in C \oplus G$ for a
variable-free term $t$ will be clear. For example, in Cohen forcing, it means
that either $t = 2n$ and $n \in C$, or $t = 2n + 1$ and $p(n) = 1$. In particular,
it will always be the case that $p \Vdash^* t \in C \oplus G$ iff $p \Vdash t \in C \oplus G$, and
that if $t \in C \oplus G$ for a $G$ corresponding to a sufficiently generic filter $F$,
then $p \Vdash^* t \in C \oplus G$ for some $p \in F$. We thus assume these properties
henceforth.

We can define $p \Vdash^* \varphi$ inductively for formulas $\varphi$ in which $G$ is the only
free variable as follows. (It is convenient to work formally with a logic in
which the only connectives are $\wedge$ and $\neg$, and the only quantifier is $\exists$, which
we can of course do with no loss of expressive power.)

1. $p \Vdash^* \neg\varphi$ iff $q \nVdash^* \varphi$ for all $q \preccurlyeq p$.
2. $p \Vdash^* \varphi \wedge \psi$ iff $p \Vdash^* \varphi$ and $p \Vdash^* \psi$.
3. $p \Vdash^* \exists n \, \varphi(n)$ iff for each $q \preccurlyeq p$, there are an $r \preccurlyeq q$ and an $n \in \mathbb{N}$ such
   that $r \Vdash^* \varphi(n)$.

It is easy to show that if $p \Vdash^* \varphi$ and $q \preccurlyeq p$, then $q \Vdash^* \varphi$, and that for every
$p$ and $\varphi$, there is a $q \preccurlyeq p$ such that $q \Vdash \varphi$ or $q \Vdash \neg\varphi$ (in other words, the
set of conditions that force $\varphi$ or force $\neg\varphi$ is dense).

The correspondence between truth (of global properties of $G$) and forcing (which is a local condition) is a crucial property of generic objects.

2 **Exercise 2.9.** Show by simultaneous induction on the structure of
formulas that if $\varphi$ is a formula in which $G$ is the only free variable then the
following hold.

(i) For every condition $p$, we have $p \Vdash \varphi$ iff $p \Vdash^* \varphi$.
(ii) Let $F$ be a sufficiently generic filter, and $G$ the corresponding generic
    set. Then $\varphi$ is true of $G$ iff there is a $p \in F$ such that $p \Vdash \varphi$.

# Chapter 3

# Finding Our Path: König's Lemma and Computability

In this chapter, we explore in greater detail the example mentioned in Section 1.2: the computability theoretic analysis of versions of König's Lemma. Once we have introduced the framework and some of the basic systems of reverse mathematics in the next chapter, we will be able to use the results and arguments in this chapter to provide a reverse mathematical analysis of versions of König's Lemma as well. As we will see, Weak König's Lemma plays a particularly important role in reverse mathematics.

To state König's Lemma precisely, we first need to define our terms.

**Definition 3.1.** A *tree* is a subset $T$ of $\mathbb{N}^{<\mathbb{N}}$ such that if $\sigma \in T$ then every initial segment of $\sigma$ is in $T$. A tree $T$ is *finitely branching* if for every $\sigma \in T$, there are only finitely many $n$ such that $\sigma n \in T$. A tree is *binary* if it is a subset of $2^{<\mathbb{N}}$. A *path* on a tree $T$ is an $X \in \mathbb{N}^{\mathbb{N}}$ such that $X \upharpoonright n \in T$ for all $n$. (We have been calling such paths "infinite paths", but since we will never deal with finite paths on trees, we will henceforth drop the word "infinite".) We denote the set of paths on $T$ by $[T]$.

We now have the following principles, first established by König [113].

**Definition 3.2.** *König's Lemma* (KL) is the statement that every infinite, finitely branching tree has a path. *Weak König's Lemma* (WKL) is the statement that every infinite binary tree has a path.

Before turning to the analysis of versions of König's Lemma, we explore the important computability theoretic notion of a $\Pi_1^0$ class, which is closely related to Weak König's Lemma. In the following section, we focus mainly on facts that will be useful later on in the book. For more on $\Pi_1^0$ classes, see [14–16].

## 3.1   $\Pi_1^0$ classes, basis theorems, and PA degrees

It is not difficult to see that a subset of $2^{\mathbb{N}}$ (or of $\mathbb{N}^{\mathbb{N}}$) is closed iff it is equal to $[T]$ for some tree $T$. The notion of a $\Pi_1^0$ class is an effectivization of this idea.

**Definition 3.3.** A $\Pi_1^0$ *class* is one of the form $[T]$ for some computable binary tree $[T]$.

The name comes from the fact that $P$ is a $\Pi_1^0$ class iff it can be defined in a $\Pi_1^0$ manner, which can be stated computability theoretically as follows.

|3| **Exercise 3.4.** Show that $P$ is a $\Pi_1^0$ class iff there is a computable set $S$ of binary strings such that $X \in P$ iff $\forall n\, X \upharpoonright n \in S$.

We may also define a $\Pi_1^0$ class in $\mathbb{N}^{\mathbb{N}}$ to be one of the form $[T]$ for some computable subtree of $\mathbb{N}^{<\mathbb{N}}$. If $T$ is finitely branching, then we get a notion corresponding to König's Lemma, so its differences from the above notion will be explored below. In general, though, $\Pi_1^0$ classes in $\mathbb{N}^{\mathbb{N}}$ behave very differently from $\Pi_1^0$ classes in $2^{\mathbb{N}}$, since $\mathbb{N}^{\mathbb{N}}$ is not compact (see [14–16]), so we will not discuss them here, and by a $\Pi_1^0$ class we will always mean one in $2^{\mathbb{N}}$.

Examples of $\Pi_1^0$ classes arise naturally in many areas of mathematics. We already saw one in connection with our discussion of Lindenbaum's Lemma. Here is another example.

|4| **Exercise 3.5.** Let $R$ be a computable infinite commutative ring (i.e., the domain of $R$ is computable, and so are its addition and multiplication functions). By applying a computable transformation (i.e., mapping the $n$th element of the domain of $R$, in the order of the natural numbers, to $n$), we may assume that the domain of $R$ is $\mathbb{N}$. Show that the class of prime ideals of $R$ is a $\Pi_1^0$ class. [Recall that we assume all rings have units.]

The next example gives us a good way to show that not all nonempty $\Pi_1^0$ classes have computable members.

|5| **Exercise 3.6.** Let $A$ and $B$ be disjoint sets. A *separating set* for $A$ and $B$ is a set $X$ such that $X \supseteq A$ and $X \cap B = \emptyset$.

**a.** Show that if $A$ and $B$ are disjoint c.e. sets, then the separating sets for $A$ and $B$ form a $\Pi_1^0$ class.

**b.** Show that there are disjoint c.e. sets with no computable separating set.

The above exercise yields the following result.

**Corollary 3.7** (Kreisel [114]). *There is a nonempty $\Pi_1^0$ class with no computable member. Thus there is a computable infinite binary tree with no computable path.*

$\boxed{6}$ **Exercise 3.8.** Give a direct proof of the second statement in Corollary 3.7.

We say that $P_0, P_1, \ldots$ are *uniformly $\Pi_1^0$ classes* if there are uniformly computable trees $T_0, T_1, \ldots$ such that $P_n = [T_n]$ for all $n$. The following fact is often useful.

**Proposition 3.9.** *For uniformly $\Pi_1^0$ classes $P_0, P_1, \ldots$, the set $\{n : P_n = \emptyset\}$ is c.e.*

*Proof.* Let $T_0, T_1, \ldots$ be uniformly computable trees such that $P_n = [T_n]$ for all $n$. Then $P_n = \emptyset$ iff $T_n$ is finite, which happens iff there is a $k$ such that $\sigma \notin T_n$ for all binary strings $\sigma$ of length $k$. The latter is a $\Sigma_1^0$ property, and hence determines a c.e. set. $\qquad\square$

We can of course relativize the notion of $\Pi_1^0$ class. A $\Pi_1^0$ class relative to $X$ (i.e., the set of paths through an $X$-computable binary tree) is called a $\Pi_1^{0,X}$ *class.*

A *basis theorem* for $\Pi_1^0$ classes is one stating that, for some class $\mathcal{C}$, every nonempty $\Pi_1^0$ class has a member in $\mathcal{C}$. The best known basis theorem is the *low basis theorem.* (Recall that a set $X$ is low if $X' \leqslant_T \emptyset'$, where $X'$ is the halting problem relativized to $X$.)

**Theorem 3.10** (Jockusch and Soare [105]). *Every nonempty $\Pi_1^0$ class has a low member.*

Proofs of this theorem can be found in [40,196], or by doing the following exercise.

$\boxed{7}$ **Exercise 3.11** (Jockusch and Soare [105]). Prove the low basis theorem using effective forcing (relative to $\emptyset'$). [Hint: Use the notion of forcing consisting of all nonempty $\Pi_1^0$ classes, ordered by inclusion. The relevant dense sets are: for each $n$, the collection of nonempty $\Pi_1^0$ classes $P$ such that for some $\sigma$ of length $n$, every element of $P$ extends $\sigma$; and for each $e$, the collection of nonempty $\Pi_1^0$ classes $P$ such that either $e \in X'$ for every $X \in P$ or $e \notin X'$ for every $X \in P$. A generic set satisfying the theorem is obtained as the intersection of all the elements of a generic filter containing the given $\Pi_1^0$ class.]

Thus every computable infinite binary tree $T$ has a low path $P$. The low basis theorem is effective, in the sense that from an index for $T$, we can computably obtain a lowness index for such a $P$ (which recall is an $e$ such that $\Phi_e^{\emptyset'} = P'$). The following extension of this fact will be useful below.

8  **Exercise 3.12** (Jockusch and Soare [105]). Show that from a lowness index for a low infinite binary tree $T$ we can computably obtain a lowness index for a path on $T$.

Lesser known but also of interest are the *hyperimmune-free basis theorem* and the *cone-avoidance basis theorem*. A set $X$ is of *hyperimmune-free degree* if for every $f \leqslant_T X$, there is a computable $g$ such that $f(n) < g(n)$ for all $n$. This is a very natural and useful notion, capturing a level of computational power insufficient to compute any very fast growing functions. One way to see that there are noncomputable sets that are "close to computable" in this is sense is via the following basis theorem.

**Theorem 3.13** (Jockusch and Soare [105]). *Every nonempty $\Pi_1^0$ class has a member of hyperimmune-free degree.*

See Soare [196] for (a sketch of) a proof, an explanation of the name "hyperimmune-free", and the fact that the only $\Delta_2^0$ sets of hyperimmune-free degree are the computable ones, though there are continuum many hyperimmune-free degrees. The hyperimmune-free degrees are also obviously closed downwards, so if a noncomputable set is $\Delta_2^0$, then it cannot be computed by any set of hyperimmune-free degree. On the other hand, if a set is not $\Delta_2^0$ then it cannot be computed by any low set, so the low basis and hyperimmune-free basis theorems together give us the following result, which can also be proved directly using a forcing construction.

**Theorem 3.14** (Jockusch and Soare [105]). *If $X$ is noncomputable then every nonempty $\Pi_1^0$ class has a member that does not compute $X$.*

The forcing proof of Theorem 3.14 shows in fact that if $X_0, X_1, \ldots$ are noncomputable then every nonempty $\Pi_1^0$ class has a member $A$ such that $X_i \not\leqslant_T A$ for all $i$. An earlier version of this result, with "$X_i \not\leqslant_T A$" replaced by "$X_i$ is not primitive recursive in $A$" appears in Gandy, Kreisel, and Tait [69].

One of the most important classes of degrees in computability theory and its applications is that of the PA degrees. The following is one of several ways to define this class.

**Definition 3.15.** A degree is a *PA degree* if the sets computable from it form a basis for the $\Pi_1^0$ classes.

In other words, $X$ has PA degree if every computable infinite binary tree has an $X$-computable path. The name for this class of degrees comes from the following theorem, which built on work of Scott [176]; a proof can be found in Odifreddi [158]. (See Section 4.1 below for the definition of Peano Arithmetic.)

**Theorem 3.16** (Solovay, see Odifreddi [158]). *A degree is PA iff it is the degree of a complete extension of Peano Arithmetic.*

It is easy to see that the class of complete extensions of a given computably axiomatizable first order theory $T$ can be coded as a $\Pi_1^0$ class. That is, given an effective listing $\varphi_0, \varphi_1, \ldots$ of the sentences in the given language, the sets $X$ such that $\{\varphi_n : n \in X\}$ is a complete extension of $T$ form a $\Pi_1^0$ class. (Cf. the discussion of Lindenbaum's Lemma in Section 1.2.) The $\Pi_1^0$ class $P$ corresponding to the completions of PA is then a *universal* $\Pi_1^0$ class, in the sense that any set that can compute an element of $P$ can compute an element of any nonempty $\Pi_1^0$ class. (Jockusch and Soare [105, 106] showed that for every $\Pi_1^0$ class $P$ there is a computably axiomatizable first order theory $T$ such that the class of degrees of members of $P$ coincides with the class of degrees of complete extensions of $T$. This result was extended by Hanf [76] to finitely axiomatizable theories.)

We will also be interested in the notion of a set $Y$ having PA degree relative to another set $X$, which means that $Y$ can compute a path on any $X$-computable infinite binary tree. This relation is denoted by $Y \gg X$. Note that if $Y \gg X$ then $Y \geqslant_T X$.

The following is one of the most important applications of the low basis theorem.

**Theorem 3.17** (Jockusch and Soare [105]). *There is a low PA degree.*

The PA degrees are clearly upwards closed, so $\emptyset'$ has PA degree. Conversely, we have the following theorem.

**Theorem 3.18** (Jockusch and Soare [106]). *If a c.e. set has PA degree then it is complete (i.e., Turing equivalent to $\emptyset'$).*

Another way to characterize PA degrees is via effectively inseparable pairs of c.e. sets.

**Definition 3.19.** The disjoint c.e. sets $A$ and $B$ are *effectively inseparable* if there is a computable function $f$ such that for all disjoint c.e. $W_e \supseteq A$ and $W_j \supseteq B$, we have $f(e, j) \notin W_e \cup W_j$.

One example of an effectively inseparable pair of sets is $A = \{e : \Phi_e(e)\!\downarrow = 0\}$ and $B = \{e : \Phi_e(e)\!\downarrow = 1\}$. It is not difficult to build a function $f$ as in the above definition using the Recursion Theorem 2.1. By Exercise 3.6, the separating sets for $A$ and $B$ form a $\Pi_1^0$ class. This $\Pi_1^0$ class is in fact universal.

**Theorem 3.20** (Jockusch and Soare [105]). *A degree is PA iff it computes a separating set for an effectively inseparable pair.*

Let $\varphi(e) = 1 - \Phi_e(e)$ if $\Phi_e(e)\!\downarrow \in \{0, 1\}$, and $\varphi(e)\!\uparrow$ otherwise. Then $\varphi$ is a partial computable $0, 1$-valued function, and any extension of $\varphi$ to a total $0, 1$-valued function is the characteristic function of a separating set for the effectively inseparable sets $A$ and $B$ defined above. Thus, any set computing such an extension has PA degree. Conversely, for any partial computable $0, 1$-valued function $\psi$, it is easy to check that the class of all total $0, 1$-valued extensions of $\psi$ is a $\Pi_1^0$ class, and hence any set of PA degree can compute such an extension. Thus we have the following characterization.

**Theorem 3.21** (Jockusch and Soare [105]). *A degree is PA iff for each partial computable $0, 1$-valued function $\psi$, it computes a total $0, 1$-valued extension of $\psi$.*

A (total) function $f$ is *diagonally noncomputable (DNC)* if $f(e) \neq \Phi_e(e)$ for all $e$. Let $\varphi$ be as above. Then a $0, 1$-valued function $f$ is DNC iff it is a total $0, 1$-valued extension of $\varphi$. Thus a degree is PA iff it computes a $0, 1$-valued DNC function. Without the condition of being $0, 1$-valued, however, there are DNC functions that are not of PA degree. DNC functions have proved to be quite important in applications of computability theory; see for instance Downey and Hirschfeldt [40].

We finish this section with a theorem of Jockusch [98] that deserves to be better known.

**Definition 3.22.** A class of sets $\mathcal{C}$ is *$X$-uniform* if it has a uniformly $X$-computable listing. A class of sets $\mathcal{C}$ is *$X$-subuniform* if there is an $X$-uniform class of sets $\mathcal{D}$ such that $\mathcal{C} \subseteq \mathcal{D}$.

Jockusch [98] showed that the class of computable sets is $X$-uniform iff $X$ is high. Lesser known is the following surprising equivalence.

**Theorem 3.23** (Jockusch [98]). *The following are equivalent.*

1. *The class of computable sets is $X$-subuniform.*
2. *$X$ is either high or has PA degree.*

## 3.2 Versions of König's Lemma

König's Lemma is easy to prove: Let $T$ be an infinite, finitely branching tree. For $\sigma \in T$, let $T_\sigma$ be the set of extensions of $\sigma$ in $T$. We define a sequence of strings by recursion. Let $\sigma_0$ be the empty string, and note that $T_{\sigma_0} = T$ is infinite. Suppose we have defined $\sigma_n$ such that $T_{\sigma_n}$ is infinite. Since $T$ is finitely branching, there must be some $k$ such that $T_{\sigma_n k}$ is infinite. Let $\sigma_{n+1} = \sigma_n k$ for the least such $k$. Now let $X = \lim_n \sigma_n$. Then $X$ is a path on $T$.

However, this proof is clearly not effective, as it requires us to be able to find strings $\sigma \in T$ such that $T_\sigma$ is infinite. Indeed, as we have seen in Corollary 3.7, even Weak König's Lemma is not computably true. On the other hand, the basis theorems of the previous section show us that computable instances of WKL always have solutions that are "close to computable" (e.g., low, of hyperimmune-free degree, cone-avoiding). The situation for the full version of König's Lemma is quite different. The following fact is proved in Jockusch, Lewis, and Remmel [103], where it is referred to as a "known result", similar to one in Yates [210].

9 **Exercise 3.24.** Show that there is a computable finitely branching tree $T$ such that $[T]$ is a singleton, and the unique element of $[T]$ computes $\emptyset'$. [Hint: Build $T$ so that, for the unique path $P$ on $T$, we have $P(n) = 0$ iff $n \notin \emptyset'$.]

For the computable instance $P$ of KL in the above exercise, there is no way to obtain a computable instance $Q$ of WKL such that any solution to $Q$ computes a solution to $P$. Thus, in the notation of Section 2.2, WKL $\leqslant_c$ KL (trivially, since WKL is a special case of KL), but KL $\not\leqslant_c$ WKL. In such a situation, it is usually of interest to try to understand more deeply the cause for the difference between these two similar principles.

In our case, a first thing we might notice is that nothing changes if we replace $2^{<\mathbb{N}}$ in WKL by $n^{<\mathbb{N}}$ for any natural number $n$. It is easy to see, for instance, that the statement that every infinite tree in $4^{<\mathbb{N}}$ has a path is equivalent to WKL. On the one hand, WKL is a special case of this statement. On the other hand, we can easily represent $\sigma \in 4^{<\mathbb{N}}$ by the string $\widehat{\sigma} \in 2^{<\mathbb{N}}$ of length $2|\sigma|$ such that for all $n < \sigma$, we have $\sigma(n) = \widehat{\sigma}(2n) + 2\widehat{\sigma}(2n+1)$ (so, for example, 2301 is represented by 10110001). Given a tree $T \subseteq 4^{<\mathbb{N}}$, let $\widehat{T} \subseteq 2^{<\mathbb{N}}$ be the tree consisting of all strings of the form $\widehat{\sigma}$ or $\widehat{\sigma}i$ (where $i \in \{0,1\}$) for $\sigma \in T$. From a path on $\widehat{T}$, we can easily obtain a path on $T$. A similar encoding works for any $n^{<\mathbb{N}}$.

Next, we might realize that we can change the encoding at each step to deal with trees where the branching grows as we go along, for instance a tree $T \subseteq \mathbb{N}^{<\mathbb{N}}$ such that for all $\sigma \in T$ and all $n < |\sigma|$, we have $\sigma(n) \leqslant n$. Considering how far we can take this idea leads us to the following result. For $S \subseteq \mathbb{N}^{\mathbb{N}}$, let $\deg(S)$ be the set of Turing degrees of elements of $S$.

$\boxed{10}$ **Exercise 3.25** (Jockusch and Soare [105, 106])**.** A tree $T$ is *computably bounded* if there is a computable function $f : \mathbb{N} \to \mathbb{N}$ such that if $\sigma \in T$ then $\sigma(n) \leqslant f(n)$ for all $n < |\sigma|$. Show that if $T$ is a computable, computably bounded tree, then there is a binary tree $\widehat{T}$ such that $\deg([\widehat{T}]) = \deg([T])$.

Thus we see that the fundamental issue here is that, in translating König's Lemma to the computability theoretic setting, we translated "infinite tree" to "computable infinite tree", but we did not translate "finitely branching" to "finitely branching in a computable manner", i.e., "computably bounded". It is the presence of a computable bounding function that explains the decreased computability theoretic complexity of WKL relative to KL.

It is worth noting that this decrease in complexity is quite pronounced. (We will see below that it leads to a real qualitative difference between the two principles in the setting of reverse mathematics.) In particular, we can think of it in terms of the idea of "coding power". Exercise 3.24 tells us that we can encode the halting problem (and hence any c.e., or even $\Delta_2^0$, set) into a computable instance of KL, in such a way that the full information of $\emptyset'$ can be decoded from any solution to that instance. Theorem 3.14 (the cone-avoidance basis theorem), on the other hand, shows that there is *no* noncomputable set that can be encoded into a computable instance of WKL so that the set is recoverable from any solution to this instance.

The following exercise shows that the coding power of KL is exactly at the level of $\emptyset'$.

11 **Exercise 3.26** (Jockusch and Soare [105, 106]). Use relativized versions of Exercise 3.25 and the cone-avoidance basis theorem to show that for every computable infinite, finitely branching tree $T$ and every $X \not\leq_T \emptyset'$, there is a path $P$ on $T$ such that $X \not\leq_T \emptyset' \oplus P$. [Hint: In relativizing Exercise 3.25, note that every computable tree is $\emptyset'$-computably bounded.]

The method of the above exercise also suffices to show that there is a path $P$ on $T$ such that $P \oplus \emptyset'$ is low relative to $\emptyset'$ (that is, $(\emptyset' \oplus P)' \leq_T \emptyset''$), or of hyperimmune-free degree relative to $\emptyset'$, and indeed that for every set $X$ of PA degree relative to $\emptyset'$, there is a path $P$ on $T$ such that $P \leq_T X$.

On the other hand, KL is not $\emptyset'$-computably true. The following exercise, together with Exercise 3.6 relativized to $\emptyset'$, gives us one way to prove this fact, in a way that explains it and clarifies the relationship between WKL and KL by showing that the ability to solve computable instances of KL gives us the power to solve $\emptyset'$-computable instances of WKL. The latter fact can be seen as a converse to the relativized version of Exercise 3.25 used in Exercise 3.26. Thus we may say that, in a sense, KL behaves like WKL "one jump up". This kind of situation, where a principle can be seen as a "higher level" version of another, is yet another potentially informative phenomenon in computable mathematics.

12 **Exercise 3.27** (Jockusch, Lewis, and Remmel [103]). Show that if $T$ is a binary tree computable in $\emptyset'$, then there is a computable finitely branching tree $\widehat{T}$ such that $\deg([\widehat{T}]) = \deg([T])$. Conclude that there is a computable infinite, finitely branching tree with no $\emptyset'$-computable path.

Note that, in the above exercise, if $[T]$ is a singleton, then we can ensure that so is $[\widehat{T}]$, giving us another way to solve Exercise 3.24.

The following is another application of this exercise. A computable instance of a principle $P$ is *universal* if any solution to it computes solutions to all computable instances of $P$. We already saw an example of this phenomenon for WKL. In that case, a computable tree whose paths code the completions of PA is a universal instance. Consider the relativization of the notion of a universal $\Pi^0_1$ class to $\emptyset'$. For example, we can let $A = \{e : \Phi_e^{\emptyset'}(e){\downarrow} = 0\}$ and $B = \{e : \Phi_e^{\emptyset'}(e){\downarrow} = 1\}$ and consider the class $P$ of all separating sets for $A$ and $B$. As in the unrelativized case of the previous section, $P$ is a $\Pi^0_1$ class relative to $\emptyset'$, and every element of $P$ has PA degree relative to $\emptyset'$. Let $T$ be an $\emptyset'$-computable binary tree such that $[T] = P$. By Exercise 3.27, there is a computable finitely branching tree $\widehat{T}$ such that $\deg([\widehat{T}]) = \deg([T])$, and hence every element of $[\widehat{T}]$ has PA degree relative to $\emptyset'$. As mentioned above, every set of PA degree relative to $\emptyset'$ can compute a path on any given computable infinite, finitely branching tree, so $\widehat{T}$ is a universal instance of KL.

We have already mentioned in Section 1.2 that restricting KL to trees with no dead ends yields a principle that *is* computably true. This is a significant fact, because trees with no dead ends arise naturally in many settings. For example, consider the version of Lindenbaum's Lemma in which, instead of being given just a consistent set of sentences, we are given a consistent theory $\Gamma$ (i.e., a consistent set of sentences that is deductively closed). Then, using the notation of Section 1.2, we can form the tree $T$ of all binary strings $\sigma$ such that $\neg\theta_\sigma \notin \Gamma$. This $T$ is computable in $\Gamma$, and it is easy to see that it has no dead ends. From a path on $T$ we obtain a completion of $\Gamma$ as in Section 1.2.

Thus, from considering the difference dead ends make to the computability theoretic strength of KL, we arrive at an interesting fact about first order logic: from the computability theoretic point of view, it is more difficult to find completions of consistent sets of sentences than it is to find completions of consistent theories; indeed, the former cannot be done computably, while the latter can. As it turns out, the proof of Gödel's completeness theorem (for countable languages) can be done computably except for the use of Lindenbaum's Lemma (see Harizanov [77] for a proof of the effective completeness theorem). Thus the completeness theorem in the version for consistent theories is computably true, while in the version for consistent sets of sentences, it is at the level of Weak König's Lemma. (We will mention the reverse mathematical version of this fact in the next chapter.)

The following is another way to obtain a computably true version of König's Lemma.

**13** **Exercise 3.28** (Kreisel [115]). Show that if a binary tree $T$ has only finitely many paths, then each of these paths is $T$-computable.

Again, this version is of practical interest; there are situations in which it is the best way to show that a particular kind of object must be computable. For example, suppose $\Gamma$ is a computable consistent theory with no finite models and only finitely many countable models. Then $\Gamma$ has only finitely many completions, so the tree constructed above has only finitely many paths, and hence every completion of $\Gamma$ is computable.

We saw in Exercise 3.26 that the result in Exercise 3.28 cannot be extended to finitely branching subtrees of $\mathbb{N}^{<\mathbb{N}}$, but it is still the case that knowing that a subtree of $\mathbb{N}^{<\mathbb{N}}$ has only finitely many paths helps us find such paths. It is not difficult to transform the solution to Exercise 3.28 into a proof that if a computable finitely branching tree has only finitely many paths, then each of these paths is $\emptyset'$-computable.

There is a difference between the two computably valid versions of König's Lemma discussed above. There is a single algorithm (the one given in Section 1.2), which, given any tree $T$ with no dead ends as an oracle, produces a path on $T$. There is, however, no single algorithm that, given any infinite binary tree $T$ with only finitely many paths as an oracle, produces a path on $T$, even if we restrict the inputs to computable infinite binary trees. To see that this is the case, suppose there is such an algorithm $P$, and build a computable binary tree $T$ as follows. Begin putting the strings of the form $0^s$ and $1^s$ into $T$, and no others, and simulate the action of $P$ on oracle $T$. (In greater detail, we proceed in stages. We may assume that, at step $s$ of its execution, $P$ does not query its oracle on any string of length greater than $s$. At each stage $s$, we declare $0^s$ and $1^s$ to be in $T$, and all other strings of length $s$ to be out of $T$, and simulate one more step in the execution of $P^T$.) Eventually, $P^T$ must converge on input 0, to some $i \in \{0, 1\}$, since otherwise $T$ is a computable infinite binary tree with only two paths but $P^T$ is not a path on $T$. If this happens at stage $s$, we then put $(1-i)^t$ into $T$ for each $t > s$, but keep all other strings of that length, including $i^t$, out of $T$. Then $T$ is a computable infinite binary tree with only one path, but $P^T$ is not a path on $T$. Thus the computable validity of the first version of KL is uniform, while that of the second is not.

In this chapter, we have seen some relatively simple examples of how a computability theoretic analysis can highlight the differences between similar principles, as well as help reveal similarities between principles arising in different areas of mathematics. We will see more such examples once we have introduced the complementary viewpoint of reverse mathematics.

# Chapter 4

# Gauging Our Strength: Reverse Mathematics

The reverse mathematics program was initiated by Friedman [59, 60] and further developed by Simpson, his students and collaborators, and, more recently, an increasing number of researchers from computability theory, proof theory, and other areas of mathematical logic. In this chapter, we formally introduce its framework and some of its basic systems. For more details and discussion, see Simpson [191]. Although we work in what is called "second order arithmetic", the underlying logic of our formal systems is the usual first order logic of any of the standard introductory textbooks in the area such as Enderton [53]. Almost all of this book will focus on principles provable in the system $\mathsf{ACA}_0$, which will be introduced in Section 4.3 and corresponds to arithmetic mathematics (i.e., mathematics that can be done without using sets outside the arithmetic hierarchy), so we will concentrate on this and weaker systems. There is a great deal of work in reverse mathematics beyond $\mathsf{ACA}_0$, however, a small amount of which we will discuss in Section 10.3.

The language of second order arithmetic is a two-sorted language with number variables (which we will denote by lowercase letters), set variables (which we will denote by uppercase letters), equality, and the usual symbols $+, \cdot, \leqslant, 0, 1$, and $\in$. Our atomic formulas are those of the form $t = u, t \leqslant u$, and $t \in X$, where $t$ and $u$ are first order terms (i.e., ones built up out of the constants, number variables, $+$, and $\cdot$ in the usual way). Formulas are then built up as usual. Note that, while $X = Y$ is not a formula, it can be expressed by $\forall n \, [n \in X \leftrightarrow n \in Y]$.

Of course, most of the principles we wish to study involve more kinds of mathematical objects than just natural numbers and sets of natural numbers. When dealing with such objects in second order arithmetic, we have to employ codings. We will postpone the discussion of this issue until

41

Section 4.1.

It is useful to allow *bounded quantifiers*, i.e., quantifiers of the form $\forall x < t$ or $\exists x < t$, where $t$ is a term in which $x$ does not appear. We may think of $\forall x < t\,\varphi$ as an abbreviation for $\forall x\,[x < t \rightarrow \varphi]$, and of $\exists x < t\,\varphi$ as an abbreviation for $\exists x\,[x < t \wedge \varphi]$.

An *arithmetic formula* is one with no quantification over set variables. Note that an arithmetic formula may contain free set variables. A *bounded quantifier formula* is an arithmetic formula in which all quantifiers are bounded. A $\Sigma_n^0$ *formula* is one of the form $\exists x_1 \forall x_2 \exists x_3 \forall x_4 \cdots Q x_n\,\varphi$, where $\varphi$ is a bounded quantifier formula, and $Q$ is $\exists$ if $n$ is odd and $\forall$ if $n$ is even. A $\Pi_n^0$ *formula* is one of the form $\forall x_1 \exists x_2 \forall x_3 \exists x_4 \cdots Q x_n\,\varphi$, where $\varphi$ is a bounded quantifier formula, and $Q$ is $\forall$ if $n$ is odd and $\exists$ if $n$ is even. Note that bounded quantifier formulas are both $\Sigma_0^0$ and $\Pi_0^0$.

The similarity with the notation for the levels of the arithmetic hierarchy in Section 2.1 is of course suggestive. Indeed, the name for that hierarchy comes from the fact that a set of natural numbers is $\Sigma_n^0$ iff it is definable in the natural numbers by a $\Sigma_n^0$ formula, and similarly for $\Pi_n^0$ sets (see e.g. [53]).

A $\Sigma_n^1$ *formula* is one of the form $\exists X_1 \forall X_2 \exists X_3 \forall X_4 \cdots Q X_n\,\varphi$, where $\varphi$ is arithmetic, and $Q$ is $\exists$ if $n$ is odd and $\forall$ if $n$ is even. A $\Pi_n^1$ *formula* is one of the form $\forall X_1 \exists X_2 \forall X_3 \exists X_4 \cdots Q X_n\,\varphi$, where $\varphi$ is arithmetic, and $Q$ is $\forall$ if $n$ is odd and $\exists$ if $n$ is even.

A *sentence* is a formula with no free variables.

A *structure* or *model* in the language of second order arithmetic is a structure of the form $(N, S, +_N, \cdot_N, 0_N, 1_N, \leqslant_N)$, where $(N, +_N, \cdot_N, 0_N, 1_N, \leqslant_N)$ is a structure in the language of first order arithmetic and $S \subseteq \mathcal{P}(N)$ (where $\mathcal{P}(N)$ is the power set of $N$). We always interpret the $\in$ symbol as actual membership.

In the following definition and the rest of this book, when we write $\varphi(n)$ for a formula $\varphi$, we assume we have distinguished some free number variable of $\varphi$, and are substituting $n$ in for that variable; of course, if $\varphi$ has no free number variables, then $\varphi(n)$ is just $\varphi$.

**Definition 4.1.** Let $P_0^-$ be the (first order) axioms for a discrete ordered commutative semiring, i.e., the universal closures of the following axioms:

$$x + y = y + x \qquad x \cdot y = y \cdot x$$

$$(x + y) + z = x + (y + z) \qquad (x \cdot y) \cdot z = x \cdot (y \cdot z)$$

$$x \cdot (y + z) = x \cdot y + x \cdot z$$

$$x + 0 = x \qquad x \cdot 0 = 0 \qquad x \cdot 1 = x$$

$$\neg(x < x) \qquad (x < y \wedge y < z) \rightarrow x < z \qquad x < y \vee y < z \vee x = y$$

$$(x < y) \rightarrow (x + z < y + z) \qquad (0 < z \wedge x < y) \rightarrow (x \cdot z < y \cdot z)$$

$$x < y \rightarrow \exists z \, (x + z = y)$$

$$0 < 1 \qquad 0 \leqslant x \qquad 0 < x \rightarrow 1 \leqslant x.$$

Let $P_0$ be $P_0^-$ together with *set induction*:

$$(0 \in X \wedge \forall n \, [n \in X \rightarrow n + 1 \in X]) \rightarrow \forall n \, [n \in X].$$

Full second order arithmetic consists of $P_0$ together with the *full comprehension scheme*

$$\exists X \, \forall n \, [n \in X \leftrightarrow \varphi(n)] \tag{4.1}$$

for each formula $\varphi$ in the language of second order arithmetic such that $X$ is not free in $\varphi$. Combining full comprehension with set induction yields the *full induction scheme*

$$(\varphi(0) \wedge \forall n \, [\varphi(n) \rightarrow \varphi(n+1)]) \rightarrow \forall n \, \varphi(n) \tag{4.2}$$

for each formula $\varphi$ in the language of second order arithmetic.

The *standard model of (first order) arithmetic* is of course the usual structure of the natural numbers. Any other model $\mathcal{N}$ of $P_0^-$ is called *nonstandard*. Such a model must contain a copy of the standard model as an initial segment. Elements of this initial segment are called *standard*, and other elements of the domain of $\mathcal{N}$ are called *nonstandard*. See Hájek and Pudlák [75] for more on nonstandard models of fragments of Peano Arithmetic.

One way to obtain subsystems of second order arithmetic is to restrict the comprehension and induction schemes.

**Definition 4.2.** For a collection of formulas $\Gamma$, the $\Gamma$-*comprehension scheme* consists of (4.1) for each formula $\varphi \in \Gamma$ such that $X$ is not free in $\varphi$, and the $\Gamma$-*induction scheme* consists of (4.2) for each formula $\varphi \in \Gamma$.

We will also have occasion to consider $\Delta_1^0$-*comprehension*, which consists of

$$\forall n \, [\varphi(n) \leftrightarrow \psi(n)] \rightarrow \exists X \, \forall n \, [n \in X \leftrightarrow \varphi(n)] \tag{4.3}$$

for each pair of formulas $\varphi, \psi$ such that $\varphi$ is $\Sigma_1^0$ and $\psi$ is $\Pi_1^0$, and $X$ is not free in $\varphi$; and $\Delta_n^0$-*induction*, which consists of

$$\forall n \, [\varphi(n) \leftrightarrow \psi(n)] \rightarrow ((\varphi(0) \wedge \forall n \, [\varphi(n) \rightarrow \varphi(n+1)]) \rightarrow \forall n \, \varphi(n))$$

for each pair of formulas $\varphi, \psi$ such that $\varphi$ is $\Sigma_n^0$ and $\psi$ is $\Pi_n^0$.

Sometimes fragments of second order arithmetic are developed with a weaker set of basic first order axioms than our $P_0^-$, as in Simpson [191]. In the presence of a small amount of induction, this choice makes no difference. (The scheme (4.2) restricted to bounded quantifier formulas $\varphi$ in the language of first order arithmetic suffices.) See the proof of Lemma II.2.1 in Simpson [191].

## 4.1   RCA$_0$

Our weak base system for reverse mathematics is known as RCA$_0$, which is pronounced "r.c.a.-nought" and stands for *Recursive Comprehension Axiom*; the subscript in this and other subsystems of second order arithmetic are there for historical reasons, to distinguish them from earlier versions with unrestricted induction. Here "recursive" is synonymous with "computable". There has been a widespread (though not universal) move in computability theory to replace "recursive", "recursively enumerable", etc. with "computable", "computably enumerable", etc. There has been no attempt to change the acronym "RCA", however.

**Definition 4.3.** The system RCA$_0$ consists of $P_0$ together with $\Delta_1^0$-comprehension and $\Sigma_1^0$-induction.

If $\varphi$ is a bounded quantifier formula, then by adding a dummy quantifier in front of $\varphi$, we can think of $\varphi$ as both a $\Sigma_1^0$ and a $\Pi_1^0$ formula (i.e., if $x$ does not occur free in $\varphi$ then both $\exists x\, \varphi$ and $\forall x\, \varphi$ are logically equivalent to $\varphi$). Thus $\Delta_1^0$-comprehension implies comprehension for bounded quantifier formulas, i.e., $\Sigma_0^0$-comprehension.

It is important to notice that, here and in other systems described below, the formulas in our comprehension and induction schemes may have parameters, i.e., free variables. For example, in RCA$_0$ we can show that the join $A \oplus B$ of two sets $A$ and $B$ exists, by applying $\Sigma_0^0$-comprehension to the formula

$$\exists m < n\,[(m \in A \land n = m + m) \lor (m \in B \land n = m + m + 1)].$$

Notice also the expression "$A \oplus B$ exists". This phrase is merely shorthand for the formula

$$\exists X\, \forall n\,[n \in X \leftrightarrow \exists m < n\,[(m \in A \land n = m + m)$$
$$\lor\ (m \in B \land n = m + m + 1)]].$$

This manner of speaking is frequently employed when discussing facts provable in subsystems of second order arithmetic.

Practically, limiting induction fits in with the idea of having as weak a reasonable base system as possible, and allows us to make certain fine distinctions that could not be made over a base system with full induction, and study notions such as the first order consequences of second order principles; we will see several examples below. More foundationally, as we will see in Exercise 4.31 below, strong induction principles can be seen as strong bounded comprehension principles, and hence it is natural to exclude them when defining systems in which we intend to have only weak set existence axioms.

Thus the presence of $\Sigma_1^0$-induction (as opposed to just $\Delta_1^0$-induction, which follows from $\Delta_1^0$-comprehension and set induction) in our base system may seem strange, particularly if we are thinking of $RCA_0$ as roughly corresponding to computable mathematics (given that $\Sigma_1^0$ properties correspond to c.e. sets rather than computable ones). It is possible to work in weaker systems such as $RCA_*$, which is defined in the same way as $RCA_0$, but with $\Sigma_1^0$-induction replaced by $\Sigma_0^0$-induction, a symbol for exponentiation added to the language, and the rules for exponentiation added to our basic first order axioms. But that choice causes many difficulties, and it is also possible to see $\Sigma_1^0$-induction as a $\Delta_0^0$ recursion principle, and hence fully in line with the correspondence between $RCA_0$ and computable mathematics, as we now discuss. (Of course, working over $RCA_*$, or other systems weaker than $RCA_0$, can reveal subtle distinctions not apparent while working over $RCA_0$, and is certainly a line of research worth pursuing.)

Consider the principle, called $PREC_1$ in Hirschfeldt and Shore [88], which states that if $\varphi(x, y)$ is a bounded quantifier formula defining a total function, then for each $z$ and $n$, there is a sequence $x_0, \ldots, x_n$ such that $x_0 = z$ and $\varphi(x_i, x_{i+1})$ holds for all $i < n$. It is easy to see that, over $P_0 + \Delta_1^0$-comprehension, this principle is equivalent to another form of recursion, the principle PREC stating that $\forall z \, \forall f \, \exists g \, [g(0) = z \land \forall n \, [g(n + 1) = f(g(n))]]$, where $f$ and $g$ range over functions (with the understanding that this formula is an abbreviation for the translation into a formula with quantification over sets, as discussed at the end of this section). As noted in [88], $PREC_1$ is equivalent to $\Sigma_1^0$-induction over $P_0 + \Delta_1^0$-induction, so either PREC or $PREC_1$ could be adopted in place of $\Sigma_1^0$-induction in the definition of $RCA_0$. These principles assert only the existence of (even just finite) iterations of given functions, and make no additional induction assumptions. Indeed, this was the route followed by Friedman [60] to define his EFT (elementary theory of functions) as the base theory to which $\Delta_1^0$-comprehension is added to get $RCA_0$.

It is not immediately obvious that $\Sigma_1^0$-induction also yields $\Pi_1^0$-induction, but the latter does hold in $RCA_0$, as does a bounded version of $\Sigma_1^0$-comprehension. We will establish these facts in Exercise 4.7.

*Peano Arithmetic* (PA) is the first order system consisting of the basic first order axioms $P_0^-$ from Definition 4.1 together with the induction scheme (4.2) for all first order formulas. The system $\Sigma_1^0$-PA is obtained from PA by restricting induction to $\Sigma_1^0$ first order formulas. We will see in Section 7.1 that $\Sigma_1^0$-PA is the first order part of $RCA_0$, i.e., that a first order statement is provable in $RCA_0$ iff it is provable in $\Sigma_1^0$-PA. (This fact, and the corresponding ones about the first order parts of the systems $ACA_0$ and $WKL_0$ discussed below, have important foundational implications. See for instance Simpson [187, 190, 191].)

A reasonable amount of mathematics can be developed in $RCA_0$. In addition to basic combinatorial facts of a clearly finitistic nature, theorems provable in $RCA_0$ include the intermediate value theorem, Urysohn's Lemma for complete separable metric spaces, the existence of algebraic closures for countable fields, and the soundness and completeness theorems of first order logic (although in the case of the completeness theorem, we need to start with a consistent theory rather than just a consistent set of sentences, as discussed in Section 3.2). This list is drawn from a longer list in Simpson [191], where proofs in $RCA_0$ of these and other examples can be found, as well as the relevant references.

Theorems like the ones in the previous paragraph deal with several kinds of mathematical objects, beyond just natural numbers and sets of natural numbers. Even in the setting of combinatorial principles, where we do not have to worry about objects like the real numbers, we still want to be able to talk about strings, trees, and so on. As mentioned above, when dealing with such objects, we have to employ codings. Fortunately, for the kinds of objects we are studying here, the process is straightforward, although working out the full formal details can take some work. We will outline the ideas here; a much more detailed account can be found in [191].

We can think of the ordered pair $\langle m, n \rangle$ as the number $(m + n)^2 + m$. Basic properties of this encoding, such as $\langle m, n \rangle = \langle m', n' \rangle \rightarrow m = m' \wedge n = n'$ can be proved in $RCA_0$. Then, when we discuss a statement such as $\forall m \, \forall n \, m \leqslant \langle m, n \rangle$ in the context of second order arithmetic, we think of it as an abbreviation, in this case for the formula $\forall m \, \forall n \, m \leqslant ((m + n) \cdot (m + n)) + m$. A function $f : \mathbb{N} \to \mathbb{N}$ is then just a set such that every element of

$f$ is an ordered pair of natural numbers, and for each $m$ there is a unique $n$ such that $\langle m, n \rangle \in f$. This definition allows us for instance to quantify over functions, as it is easy to transform a statement involving quantification over formulas into one involving quantification over sets. Similarly, we think of a family of sets $S_0, S_1, \ldots$ as a single set $S = \{\langle i, n \rangle : n \in S_i\}$. It is important to note that when we say that certain sets $S_0, S_1, \ldots$ exist, we always mean it in the sense that this set $S$ exists, not just that each individual $S_i$ exists. For example, while RCA$_0$ is of course enough to show that for each $n$, the set $\{n\}$ exists (i.e., to prove for each $n$ that there is an $X$ such that $m \in X \leftrightarrow m = n$), it is not enough to prove for each function $f$ that there are sets $S_0, S_1, \ldots$ such that $S_n = \{2n\}$ if $\exists m \, f(m) = n$ and $S_n = \{2n + 1\}$ otherwise. (This fact follows from Exercise 4.11 below.)

To each finite set $F$ we assign a unique natural number called the *code* of $F$. One way to do this is to pick this code to be the least number of the form $\langle k, \langle m, n \rangle \rangle$ such that $i \in F$ iff $i < k$ and $m(i + 1) + 1$ divides $n$. (See [191] for a proof in RCA$_0$ that such a number always exists.) A finite sequence of natural numbers $(n_0, \ldots, n_{k-1})$ can be encoded as the set of ordered pairs $\{\langle i, n_i \rangle : i < k\}$ (i.e., the function from $\{0, \ldots, k - 1\}$ to $\mathbb{N}$ taking $i$ to $n_i$). Then $\mathbb{N}^{<\mathbb{N}}$ is just the set of codes of finite sequences of natural numbers, and it is straightforward to define notions such as concatenation and prove their basic properties in RCA$_0$. An infinite sequence of natural numbers is just a function from $\mathbb{N}$ to $\mathbb{N}$. Thus an infinite sequence is a second order object, so the set $\mathbb{N}^{\mathbb{N}}$ of all such sequences would be a third order object. We do not deal with such objects directly, so we think of a statement such as $f \in \mathbb{N}^{\mathbb{N}}$ as nothing but an abbreviation for "$f$ is an infinite sequence" (which of course itself abbreviates "every element of $f$ is an ordered pair and for each $m$ there is a unique $n$ such that $\langle m, n \rangle \in f$", as discussed above). A tree is a subset of $\mathbb{N}^{<\mathbb{N}}$ that contains the code of $\{\langle i, n_i \rangle : i < k\}$ whenever it contains the code of $\{\langle i, n_i \rangle : i < l\}$ for $l > k$, and a path on a tree $T$ is an infinite sequence $f$ such that the code of $\{\langle i, f(i) \rangle : i < k\}$ is in $T$ for all $k$.

Given these definitions, we can state principles such as König's Lemma fully formally in the language of second order arithmetic (although writing them out explicitly can be a bit tedious, and we will never actually do so). There are other similar objects that we will consider below, such as orderings, for example; the details of formalization for such objects are left as exercises.

We could of course choose different reasonable coding schemes, but the main thing to notice is that our codings are "transparent", in the sense that any two reasonable choices lead to equivalent statements (over $\mathrm{RCA}_0$), so in practice we can forget about them. When studying other areas from a reverse mathematical perspective, the choice of codings can make a difference. This phenomenon is particularly noticeable in areas such as analysis (or, even more problematically, topology), which, because they deal with uncountable objects, can traditionally be approached reverse mathematically only with the use of "good enough" countable surrogates, such as countable dense subsets of separable metric spaces. See [191] for more on this subject. A proposal for a version of reverse mathematics that does away with the need for coding has been made by Friedman [62]. In this *strict reverse mathematics*, there should be a separate theory for each area of mathematics, with the basic notions of that area taken as primitives, and logical axioms replaced by natural mathematical statements.

## 4.2   Working in $\mathrm{RCA}_0$

In practice, we will not work directly in $\mathrm{RCA}_0$ as a formal system, but will argue informally about what can be proved in it. This standard practice is much like the usual style of argumentation in computability theory, where to show that a certain set $A$ is computable, say, we describe an algorithm for computing $A$ in natural language, which acts to convince the reader that one could explicitly define a Turing machine computing $A$. The actual definition of such a machine, or of some equivalent object in another formalism such as a particular programming language, is almost never carried out. Similarly, we will not resort to formal proofs, and in many cases the process of converting our proofs to formal ones would be lengthy and tedious, even if basically straightforward.

In particular, we do not generally bother to spell out the codings discussed in Section 4.1, for instance writing $\exists y_0, \ldots, y_{n-1} \,\forall i < n \,\theta(y_i)$ instead of the formula stating that there is an $m$ that is the code for a sequence of length $n$ all of whose elements satisfy $\theta$. One point worth noting is that $\mathrm{RCA}_0$ suffices to "push bounded quantifiers in". For example, let $\varphi \equiv \forall x < n \,\exists y \,\theta(x, y)$, where $\theta$ is a bounded quantifier formula. Then $\varphi$ is equivalent over $\mathrm{RCA}_0$ to $\exists y_0, \ldots, y_{n-1} \,\forall i < n \,\theta(i, y_i)$, so we may apply $\Sigma^0_1$-induction to $\varphi$. Again, we normally do not spell out manipulations of this sort.

It is often easiest to think that we are arguing within a given model of $\mathrm{RCA}_0$. That is, we fix a structure $\mathcal{M} = (M, S, +_M, \cdot_M, 0_M, 1_M, \leqslant_M)$

in which $\mathrm{RCA}_0$ holds, and translate all our statements into statements about that structure. For example, "$n \in \mathbb{N}$" means that $n \in M$, and $\forall f \, \exists X \, [n \in X \leftrightarrow \exists m \, f(m) = n + 1]$ means that for each $f \in S$ that is a function, there is an $X \in S$ whose elements are exactly those $n \in M$ such that $f(m) = n +_M 1_M$ for some $m \in M$. We can then reason in the same semi-formal style that we would in any other part of mathematics.

Of particular interest are the notions of finite and infinite. Let us begin with the former. We should first of all be careful to distinguish between "there are only finitely many $x$ such that $P(x)$" and "there is a finite set $F$ consisting of all $x$ such that $P(x)$", where $P$ is some property expressible in the language of second order arithmetic. The first phrase has as a formal analog the sentence $\exists x \, \forall y \, [P(y) \rightarrow y \leqslant x]$, so when interpreted in $\mathcal{M}$, it means that there is a bound $b \in M$ such that if $\mathcal{M} \vDash P(n)$ then $n \leqslant_M b$. A subset $X$ of $M$ is *bounded* if there is a $b \in M$ such that $n \leqslant_M b$ for all $n \in X$; otherwise $X$ is *unbounded*. Thus our first phrase should be interpreted as saying that the set of elements of $M$ with property $P$ is bounded. But not every subset of $M$ is necessarily in $S$; in particular, there may be bounded subsets of $M$ that are not in $S$. Indeed, if the first order part of $\mathcal{M}$ is nonstandard, then this must be the case: Let $X \subset M$ be the standard part of $M$ (i.e., the set consisting of $0_M, 1_M, 1_M +_M 1_M, 1_M +_M 1_M +_M 1_M, \ldots$). Then $X$ is bounded (with any $b \in M \setminus X$ as a bound), but $X$ cannot be in $S$, as otherwise set induction would be violated (see Definition 4.1).

Our second phrase has a couple of reasonable formal analogs. One of them is the sentence $\exists F \, [\exists x \, \forall y [F(y) \rightarrow y \leqslant x] \wedge \forall z \, [z \in F \leftrightarrow P(z)]]$ (i.e., there is a set $F$ that is bounded and consists of all $z$ such that $P(z)$). Another uses codes for finite sets, as discussed in the previous section. Let us use the particular coding scheme described there. Let $\varphi(c, z)$ be a formula stating that $c$ is of the form $\langle k, \langle m, n \rangle \rangle$, where $z < k$ and $m(z+1)+1$ divides $n$. Then we can translate our phrase as $\exists c \, \forall z \, [\varphi(c, z) \leftrightarrow P(z)]$ (i.e., there is a code for a finite set consisting of all $z$ such that $P(z)$). Fortunately, these two interpretations are equivalent for models of $\mathrm{RCA}_0$. An element of $S$ is $\mathcal{M}$-*finite* if it is coded by an element of $M$ in the sense we have been discussing.

**14** **Exercise 4.4.** Let $\mathcal{M} = (M, S, +_M, \cdot_M, 0_M, 1_M, \leqslant_M)$ be a model of $\mathrm{RCA}_0$. Show that an element of $S$ is bounded iff it is $\mathcal{M}$-finite.

In many ways, $\mathcal{M}$-finite sets behave like finite subsets of $\omega$. For instance, every nonempty $\mathcal{M}$-finite set has a least element and a greatest element. But one should also proceed with some care. As we have seen, not every bounded subset of $M$ is necessarily $\mathcal{M}$-finite, and hence a subset of an $\mathcal{M}$-finite set need not be $\mathcal{M}$-finite (that is, it may not be in $S$). In particular, it is not in general provable in $\mathrm{RCA}_0$ that if $P$ is a property and $F$ is a finite set, then $\{x \in F : P(x)\}$ is a (finite) set.

Turning now to the infinite, we again should distinguish between "there are infinitely many $x$ such that $P(x)$" and "there is an infinite set consisting of all $x$ such that $P(x)$". Here, of course, the distinction is more obvious, as going from the former statement to the latter in general requires full comprehension. In the structure $\mathcal{M}$, the former statement is translated by saying that there are unboundedly many $n \in M$ with property $P$ (i.e., for each $m \in M$, there is an $n \geqslant_M m$ with property $P$), while the latter is translated by saying that there is an unbounded element of $S$ consisting of all $n \in M$ with property $P$. We refer to an unbounded element of $S$ as $\mathcal{M}$-*infinite*.

The basic apparatus of computability theory can be developed in $\mathrm{RCA}_0$. One way to do so, using universal formulas, is described in Section VII.1 of Simpson [191]. Another is to follow the standard development of Turing machines as in Soare [196], for instance, to define the set $S$ of all $\langle e, m, s \rangle$ such that the $e$th Turing machine on input $m$ halts in at most $s$ many steps. Having fixed a listing $D_0, D_1, \ldots$ of the finite sets, we then write $\Phi_e^A(n){\downarrow}$ to mean that there are an $s$ and $i, j, x < s$ with $D_i \subseteq A$ and $D_j \subseteq \overline{A}$ such that $\langle e, \langle n, i, j, x \rangle, s \rangle \in S$. In this case, for the least such $s$ and the least corresponding triple $\langle i, j, x \rangle$ (which exist by the least number principle for $\Sigma_1^0$ formulas), we write $\Phi_e^A(n){\downarrow} = x$ and $\Phi_e^A(n)[t]{\downarrow} = x$ for all $t \geqslant s$. We then say that $B$ is computable in $A$ if there is an $e$ such that $n \in B$ iff $\Phi_e^A(n){\downarrow} = 1$ and $n \notin B$ iff $\Phi_e^A(n){\downarrow} = 0$. Similarly, $B$ is c.e. in $A$ if there is an $e$ such that $n \in B$ iff $\Phi_e^A(n){\downarrow}$. We write $B = A'$ to mean that $e \in B$ iff $\Phi_e^A(e){\downarrow}$ and $B = A'[s]$ to mean that $e \in B$ iff $\Phi_e^A(e)[s]{\downarrow}$.

It is important to note that in a model $\mathcal{M}$ of $\mathrm{RCA}_0$, all elements of the domain of the first order part of $\mathcal{M}$ are indices for functionals. That is, when we talk about $\Phi_e$ in $\mathcal{M}$, the index $e$ may be nonstandard.

One thing these definitions allow us to do is to express computability theoretic notions as principles that can be studied reverse mathematically. For example, recall that a function $f$ is diagonally noncomputable relative to $A$ if $f(e) \neq \Phi_e^A(e)$ for all $e$. The *Diagonally Nonrecursive Principle* (DNR) is the statement that for each set $A$ there is a function that is diagonally noncomputable relative to $A$. (As with $\mathrm{RCA}_0$, the acronym "DNR" in reverse mathematics has tended to remain unchanged despite the terminological shift from "recursive" to "computable".) We will see in Section 9.1

that this principle fits in an interesting way with the picture of the reverse mathematics of combinatorial principles that we will develop. In Section 9.3, we will give two examples of model theoretic principles (called OPT and AST) that are equivalent to natural computability theoretic principles over RCA$_0$.

We can also define computability theoretic notions over models of $\Sigma_1^0$-PA. Let $\mathcal{N}$ be a (first order) model of $\Sigma_1^0$-PA with domain $N$. We say that $A \subseteq N$ is c.e. if it is $\Sigma_1^0$-definable in $\mathcal{N}$, i.e., of the form $\{n \in N : \mathcal{N} \vDash \varphi(n)\}$ for some $\Sigma_1^0$ formula with parameters from $N$. We say that $A \subseteq N$ is computable if it and its complement are both c.e., i.e., if it is $\Delta_1^0$-definable in $\mathcal{N}$. Fix a listing $D_0, D_1, \ldots$ of the $\mathcal{N}$-finite sets. For $A, B \subseteq N$, we say that $B$ is c.e. in $A$ if there is a c.e. set $W \subseteq N$ such that $n \in B$ iff $\exists i\, \exists j\, [D_i \subseteq A \wedge D_j \subseteq \overline{A} \wedge \langle n, i, j \rangle \in W]$. We say that $B$ is computable in $A$ if both $B$ and its complement are c.e. in $A$. If $\mathcal{M}$ is a model of RCA$_0$ with first order part $\mathcal{N}$, then these notions coincide with the ones defined above for all $A$ and $B$ in the second order part of $\mathcal{M}$.

There are some subtleties here. For instance, if we use the above notions to define Turing reducibility for subsets of $N$ (not arguing within a model of RCA$_0$), we do not necessarily end up with a transitive relation. This issue can be fixed by saying that $B$ is strongly c.e. in $A$ if there is a c.e. set $W \subseteq N$ such that $D_k \subseteq B$ iff $\exists i\, \exists j\, [D_i \subseteq A \wedge D_j \subseteq \overline{A} \wedge \langle k, i, j \rangle \in W]$, and altering the other definitions accordingly. However, for our purposes these issues will not arise, so we may safely ignore them. For more on this subject, see Chong and Yang [25, 26]. Although its focus is different, another useful source on defining computability theoretic notions in settings with limited induction is Sacks [175]; see in particular Section VII.3.

Most people studying reverse mathematics for the first time take a while to become fully comfortable with what informal arguments are truly translatable to proofs in RCA$_0$. Indeed there are pitfalls that even experienced researchers fall into once in a while, for example the fact (which will be discussed further in Section 6.8) that it is *not* provable in RCA$_0$ that if we partition $\mathbb{N}$ into finitely many parts, at least one of these parts must be infinite.

The issue here is slightly subtle. Suppose we have sets $F_0$ and $F_1$, and we know both are finite. Then we have $(\exists x_0 \, \forall y \geqslant x_0 \, y \notin F_0) \wedge (\exists x_1 \, \forall y \geqslant x_1 \, y \notin F_1)$. From this formula it follows in RCA$_0$ that $\exists x \, \forall y \geqslant x \, [y \notin F_0 \wedge y \notin F_1]$, i.e., that $F_0 \cup F_1$ is finite. One can give a formal derivation of this implication, but it is easier to argue in a model $\mathcal{M} = (M, S, +_M, \cdot_M, 0_M, 1_M, \leqslant_M)$ of RCA$_0$. If such a model satisfies our first formula for $F_0, F_1 \in S$, then

there are $n_0, n_1 \in M$ such that for all $m \in M$, if $n_0 \leqslant_M m$ then $m \notin F_0$ and
if $n_1 \leqslant_M m$ then $m \notin F_1$. Since $\leqslant_M$ is a linear order, there is an $n \in M$
such that $n_0, n_1 \leqslant_M n$. Then for all $m \in M$, if $n \leqslant_M m$ then $m \notin F_0 \cup F_1$.
(Note that $F_0 \cup F_1 \in S$ by $\Sigma_0^0$-comprehension.)

This argument works just as well for any fixed $k \in \omega$ and finite sets
$F_0, \ldots, F_k$. But suppose now that we do not fix $k$, but try to prove the
general statement that the union of finitely many finite sets is finite. To
be precise, suppose we are given a set $F \in S$ and a $k \in M$ such that for
each $i \leqslant_M k$, we know that the set $F_i = \{n : \langle i, n \rangle \in F\}$ is $\mathcal{M}$-finite, and
we wish to show that $\bigcup_{i \leqslant k} F_i$ is $\mathcal{M}$-finite. (Or conversely, that if $\bigcup_{i \leqslant k} F_i$
is $\mathcal{M}$-infinite then at least one of the $F_i$ is $\mathcal{M}$-infinite. An example of this
situation is when we partition $M$ into sets $F_0, \ldots, F_k$.) Then our argument
runs into a problem. We have $n_0, \ldots, n_k \in M$ such that for each $i \leqslant k$,
if $n_i \leqslant_M m$ then $m \notin F_i$. But $k$ might be nonstandard, so there is no
reason to conclude that there is an $n \in M$ such that $n_0, \ldots, n_k \leqslant_M n$. In a
nonstandard model of $P_0^-$, there can be cofinal sequences of "finite" length
(i.e., unbounded sequences whose length is a nonstandard number). It is
tempting to try to prove the existence of an $n$ as above by induction, but
$\Sigma_1^0$-induction is not enough in this case. (As we will see in Section 6.8, the
relevant first order principle here is what is known as $\Sigma_2^0$-bounding, which is
strictly intermediate in strength between $\Sigma_1^0$-induction and $\Sigma_2^0$-induction.)

In learning to prove theorems within RCA$_0$, it is advisable at first
to write down arguments in detail, noting in particular exactly how $\Delta_1^0$-
comprehension and/or $\Sigma_1^0$-induction are being used, and mentioning codings explicitly. Examples and exercises can be found in Simpson [191];
relevant exercises will also be given in this and the next few sections. It is
also a good exercise to fill in the details in some of the arguments in RCA$_0$
presented in the rest of this book. A good example of how proofs that do
not quite carry through in RCA$_0$ can sometimes be modified to do so by
careful attention to details is given by the proof of Theorem 6.32 below,
and the discussion preceding it.

15  **Exercise 4.5.** Show that RCA$_0$ proves that every infinite tree with
no dead ends has a path, noting exactly how $\Delta_1^0$-comprehension and/or
$\Sigma_1^0$-induction are being used, and mentioning codings explicitly.

It is also a good idea to see fully worked out proofs in RCA$_0$ of basic principles that we employ repeatedly (and typically without comment)
when arguing less formally in RCA$_0$. Once again, Simpson [191] is an excellent source. A good example is that book's Section II.3, which establishes

the provability in RCA$_0$ of some basic properties of functions, as well as two frequently useful principles: bounded $\Sigma_1^0$-comprehension and $\Pi_1^0$-induction. The following is an outline of the results of that section.

**Theorem 4.6** (Friedman, see Simpson [191]). *It is provable in* RCA$_0$ *that the class of total functions is closed under composition, primitive recursion, and the least number operator. (See [191] for precise statements of these operations.)*

As Simpson [191] puts it, this theorem implies that "elementary number theory can be developed straightforwardly within RCA$_0$." Indeed, even weaker theories suffice to establish many arguments in elementary number theory and other areas such as finite combinatorics. See [191] for further comments on this topic.

**16** **Exercise 4.7** (Friedman, see Simpson [191]).
**a.** Show that it is provable in RCA$_0$ that for any infinite set $X$, there is an increasing function $f$ such that for each $n \in X$ there is an $m$ with $f(m) = n$.
**b.** Let $\varphi$ be a $\Sigma_1^0$ formula in which $Y$ and $g$ do not occur free. The following is provable in RCA$_0$. Either there is a finite set $Y$ such that $\forall n \, [n \in Y \leftrightarrow \varphi(n)]$ or there is an injective function $g$ such that $\forall n \, [\varphi(n) \leftrightarrow \exists m \, g(m) = n]$. [Note that this statement is an analog to the computability theoretic fact that infinite c.e. sets are ranges of injective functions.]
**c.** Show that RCA$_0$ proves the *bounded $\Sigma_1^0$-comprehension* scheme, which states that
$$\forall b \, \exists X \, \forall n \, [n \in X \leftrightarrow (n < b \wedge \varphi(n))]$$
for each $\Sigma_1^0$ formula $\varphi$ such that $X$ is not free in $\varphi$.
**d.** Show that RCA$_0$ proves the $\Pi_1^0$-induction scheme.
**e.** Show that $P_0 + \Delta_1^0$-comprehension + bounded $\Sigma_1^0$-comprehension proves $\Sigma_1^0$-induction.

While we are mostly interested in equivalences over RCA$_0$, the following facts will be useful in Section 7.3.

**17** **Exercise 4.8** (Paris and Kirby [161]).
**a.** Show that $\Sigma_1^0$-induction and $\Pi_1^0$-induction are equivalent over $P_0$.
**b.** Show that $P_0^- + \Sigma_1^0$-induction proves the $\Sigma_1^0$-*bounding* scheme, which states that
$$\forall n \, [\forall i < n \, \exists k \, \varphi(i,k) \rightarrow \exists b \, \forall i < n \, \exists k < b \, \varphi(i,k)]$$

for each $\Sigma_1^0$ formula $\varphi$ such that $b$ is not free in $\varphi$.

## 4.3 ACA$_0$

**Definition 4.9.** The system ACA$_0$ (for *Arithmetic Comprehension Axiom*) consists of $P_0$ together with the comprehension scheme (4.1) for each arithmetic formula $\varphi$ such that $X$ is not free in $\varphi$.

Note that, combined with set induction, arithmetic comprehension yields arithmetic induction, i.e., (4.2) for each arithmetic formula $\varphi$. We will see in Section 7.1 that PA is the first order part of ACA$_0$.

It might seem that there should be a hierarchy of systems between RCA$_0$ and ACA$_0$, with levels of comprehension rising up the arithmetic hierarchy, but actually we have the following fact.

18 **Exercise 4.10** (see Simpson [191]). Show that ACA$_0$ is equivalent over RCA$_0$ to $\Sigma_1^0$-comprehension (i.e., the comprehension scheme (4.1) for each $\Sigma_1^0$ formula $\varphi$ such that $X$ is not free in $\varphi$). [Hint: The formula $\varphi$ may have parameters.]

One way to think of a $\Sigma_1^0$ formula $\varphi(x)$ is that it defines a c.e. set (or, more accurately, a set that is c.e. relative to the parameters in $\varphi$). Since the c.e. sets are the ranges of computable functions, we have the following useful characterization.

19 **Exercise 4.11** (Friedman [60]).
**a.** Show that, over RCA$_0$, ACA$_0$ is equivalent to the statement that for each $f : \mathbb{N} \to \mathbb{N}$, there is an $A$ such that $\forall x\,[x \in A \leftrightarrow \exists y\,[f(y) = x]]$ (i.e., $A = \operatorname{rng} f$; informally we say that "the range of $f$ exists").
**b.** Show that this equivalence remains true if we restrict ourselves to injective functions $f$, but not if we restrict ourselves to increasing functions $f$. [Cf. the second part of Exercise 4.7.]

Let $T$ be a principle of the form "for all $X \in \mathcal{C}$, there is a $Y$ such that $R(X,Y)$." Suppose that we can prove that there is a computable $X \in \mathcal{C}$ such that if $R(X,Y)$ then $Y$ computes $\emptyset'$. Then it is often the case that this proof can be modified to show that $T$ implies ACA$_0$ over RCA$_0$, by replacing the coding of $\emptyset'$ by a coding of the range of a given function $f$ and appealing to Exercise 4.11. The following exercise provides an example of this method, applied to the solution to Exercise 3.24.

**20** **Exercise 4.12** (Friedman [59, 60]). Show that König's Lemma is equivalent to $ACA_0$ over $RCA_0$.

There are many other theorems equivalent to $ACA_0$ over $RCA_0$. We will see some below, but others include the Bolzano-Weierstraß Theorem, the existence of maximal ideals for countable commutative rings, and the existence of bases for countable vector spaces. See [190, 191] for a longer list of examples, and [191] for proofs of the equivalences of these and other theorems to $ACA_0$, as well as the relevant references.

A great deal of mathematics can be developed within $ACA_0$, but certainly not all. As mentioned above, in this book we will focus on the world below $ACA_0$, but there are many important combinatorial principles that live outside this world. We will discuss a few of these briefly in Section 10.3, in which we will also introduce $ATR_0$ and $\Pi_1^1\text{-}CA_0$, the two systems that, together with $RCA_0$, $WKL_0$ (discussed in the following section), and $ACA_0$, constitute what are usually considered the "big five" systems of reverse mathematics. The reason for the particular interest in these systems is that there is a large number of theorems, in many different areas of mathematics, each of which has been proved equivalent to one of these five systems, while no other systems with such large and diverse equivalence classes (under implication over $RCA_0$) have yet been found. (One could argue for the inclusion in this list of another system, $WWKL_0$, which will be introduced in the next section. See also the comments on the work of Montalbán on Fraïssé's Conjecture in Section 10.3.) We will say a little more on this topic in Section 4.7.


## 4.4   WKL₀

Despite Exercise 4.10, there *is* an important subsystem of second order arithmetic between $RCA_0$ and $ACA_0$, and it turns out be our old friend Weak König's Lemma.

**Definition 4.13.** The system $WKL_0$ consists of $RCA_0$ together with Weak König's Lemma.

**21** **Exercise 4.14** (Friedman, Simpson, and Smith [67]). Adapt the solution to Exercise 3.25 to show that the following statement is equivalent to $WKL_0$ over $RCA_0$: If $T$ is an infinite tree and there is a function $f$ such that for every $\sigma \in T$, we have $\sigma(n) \leqslant f(n)$ for all $n < |\sigma|$, then $T$ has a path.

Compactness arguments abound in mathematics, even if sometimes in disguise. One of the triumphs of reverse mathematics has been to give precise mathematical form to this empirical observation. Weak König's Lemma is a way to state the compactness of $2^{\mathbb{N}}$. Many other spaces can be embedded into $2^{\mathbb{N}}$ in such a way that their compactness can be proved in WKL$_0$. Not only does this compactness allow us to prove many familiar theorems, but it is in many cases *indispensable* to proving these theorems. This indispensability can be rigorously proved in the form of equivalence with WKL$_0$. One example we have already encountered is Lindenbaum's Lemma.

$\boxed{22}$ **Exercise 4.15** (Simpson [188]). Adapt the argument given in Section 1.2 to show that Lindenbaum's Lemma is equivalent to WKL$_0$ over RCA$_0$.

Other examples of theorems equivalent to WKL$_0$ over RCA$_0$ include the compactness of $[0, 1]$ (and various closely related theorems such as the fact that any continuous real-valued function on $[0, 1]$ has a supremum), the local existence theorem for solutions of ordinary differential equations, the completeness theorem for consistent sets of formulas (cf. the mention of the completeness theorem in Section 4.1), the existence of primes ideals for countable commutative rings (cf. the mention of maximal ideals in Section 4.3), and Brouwer's Fixed Point Theorem. See [190, 191] for a longer list of examples, and [191] for proofs of the equivalences of these and other theorems to WKL$_0$, as well as the relevant references.

It should not be surprising given their definitions that WKL$_0$ and PA degrees are closely connected, as will be further spelled out in the next section. For instance, the following frequently used tool for working with WKL$_0$ is a reverse mathematical analog to Theorem 3.21, since for any $\varphi$ and $\psi$ as in its statement, there is a partial computable $0, 1$-valued function $f$ such that $f(n) = 0$ iff $\varphi_0(n)$ holds and $f(n) = 1$ iff $\varphi_1(n)$ holds.

$\boxed{23}$ **Exercise 4.16** (see Simpson [191]). Show that WKL$_0$ is equivalent over RCA$_0$ to the $\Sigma^0_1$-*separation principle*, which states that for any $\Sigma^0_1$ formulas $\varphi_0(n)$ and $\varphi_1(n)$ in which $X$ does not occur free, if $\neg \exists n\,[\varphi_0(n) \wedge \varphi_1(n)]$ then

$$\exists X\,[(\varphi_0(n) \rightarrow n \notin X) \wedge (\varphi_1(n) \rightarrow n \in X)].$$

We will see in Section 7.2 that the first order part of WKL$_0$ is the same as that of RCA$_0$, i.e., $\Sigma^0_1$-PA.

In addition to the many theorems equivalent to $WKL_0$, there are also several, provable in $WKL_0$ but not in $RCA_0$, that are strictly weaker than $WKL_0$. This phenomenon is particularly prominent in the reverse mathematical analysis of measure theory, where the following weakening of $WKL_0$ has proved important.

**Definition 4.17.** The system $WWKL_0$ consists of $RCA_0$ together with *Weak Weak König's Lemma*, which states that if $T$ is a binary tree such that

$$\liminf_n \frac{|\{\sigma \in T : |\sigma| = n\}|}{2^n} > 0$$

then $T$ has a path.

Note that if a binary tree $T$ is "fat" in the sense of the above definition, then $[T]$ has positive measure, so $WWKL_0$ is a way of capturing the difficulty of the general problem of obtaining an element of a $\Pi_1^0$ class given that we know the class has many members. Section X.1 of Simpson [191] lists several theorems that are equivalent to $WWKL_0$ over $RCA_0$, for example the fact that any Borel measure on a compact metric space is countably additive. There are also important connections between $WWKL_0$ and the theory of algorithmic randomness; we will briefly mention one such connection in the next section.

## 4.5 $\omega$-Models

We still have not shown that $RCA_0$, $ACA_0$, and $WKL_0$ are all different. Notice that computability theoretic nonimplications do not immediately transfer to the setting of reverse mathematics. For example, we know that König's Lemma is equivalent to $ACA_0$ over $RCA_0$, so to separate $ACA_0$ from $WKL_0$, it is enough to show that Weak König's Lemma does not imply König's Lemma. We have seen that WKL and KL are quite different computability theoretically, and in particular that WKL does not imply KL in the computability theoretic sense. However, this fact is not quite enough to show that WKL does not imply KL over $RCA_0$. For one thing, there could be a way to obtain solutions to instances of KL not from one, but from multiple iterated applications of WKL. So how do we show that there is no proof of KL in $WKL_0$? As in many other settings, nonimplications in reverse mathematics are usually established by considering models. Of particular interest are models in which the first order part is the standard model of arithmetic.

**Definition 4.18.** An $\omega$-*model* is one whose first order part is standard. That is, it is of the form $(\omega, S, +, \cdot, 0, 1, \leqslant)$ with $S \subseteq \mathcal{P}(\omega)$. (Here we use $\omega$ to mean the standard natural numbers.) We may identify such a model with $S$, and will do so freely.

The following exercise goes a long way towards explaining the close connections between reverse mathematics and computability theory.

**24** **Exercise 4.19** (Friedman [59]).
a. Show that a nonempty $S \subseteq \mathcal{P}(\omega)$ is an $\omega$-model of $\mathrm{RCA}_0$ iff it is a Turing ideal, as defined in Section 2.2.
b. Show that a nonempty $S \subseteq \mathcal{P}(\omega)$ is an $\omega$-model of $\mathrm{ACA}_0$ iff it is a *jump ideal*, that is, a Turing ideal such that if $X \in S$ then $X' \in S$.

These facts are not specific to $\omega$-models. Let $\mathcal{N}$ be a (first order) model of $\Sigma_1^0$-PA with domain $N$. For $S \subseteq \mathcal{P}(N)$, let $(\mathcal{N}, S)$ be the structure with first order part $\mathcal{N}$ and second order part $S$. Suppose that $(\mathcal{N}, S) \models \mathrm{I}\Sigma_1^0$. Then, interpreting computability theoretic notions in the sense of $\mathcal{N}$ as discussed in Section 4.2, $(\mathcal{N}, S) \models \mathrm{RCA}_0$ iff $S$ is a Turing ideal, and $(\mathcal{N}, S) \models \mathrm{ACA}_0$ iff $S$ is a jump ideal. (Of course, in the latter case, the existence of such an $S$ implies that $\mathcal{N} \models \mathrm{PA}$.)

It follows from the above exercise that $\mathrm{RCA}_0$ and $\mathrm{ACA}_0$ each have a minimal $\omega$-model, namely the collection of all computable sets and the collection of all arithmetic sets, respectively.

Let $P$ be a principle expressed by a sentence of the form $\forall X\, [\theta(X) \rightarrow \exists Y\, \psi(X, Y)]$ with both $\theta$ and $\psi$ arithmetic. Suppose we can show that $P$ is not computably true by showing that there is a computable $X$ such that $\theta(X)$ holds but there is no computable $Y$ such that $\psi(X, Y)$ holds. Then we know that $P$ fails in the minimal $\omega$-model $\mathcal{M}$ of $\mathrm{RCA}_0$, and hence is not provable in $\mathrm{RCA}_0$. (The requirement that $\theta$ and $\psi$ be arithmetic is actually necessary. For instance, if $\theta$ is not arithmetic, then there may be computable sets $X$ such that $\theta(X)$ holds but $\mathcal{M} \nvDash \theta(X)$. See Section 10.2 for more on this issue.) For example, by Corollary 3.7, there is a computable infinite binary tree with no computable path, so we have the following result.

**Theorem 4.20** (Friedman [59]). $\mathrm{RCA}_0 \nvdash \mathrm{WKL}_0$.

25 **Exercise 4.21** (Yu and Simpson [211]). Show that $\mathrm{RCA}_0 \nvdash \mathrm{WWKL}_0$.

The $\omega$-models of $\mathrm{WKL}_0$ are called *Scott sets*. Since the completions of PA form a $\Pi_1^0$ class, every Scott set contains a set of PA degree. Scott [176] showed that for any completion $C$ of Peano Arithmetic, the sets representable in $C$ form a Scott set. (Representability here is in the usual sense of first order logic; see for instance Enderton [53].) Any set representable in $C$ is computable in $C$, so we have the following result.

**Theorem 4.22** (Scott [176]). *For every set $A$ of PA degree, there is an $\omega$-model of $\mathrm{WKL}_0$ consisting entirely of $A$-computable sets.*

By Theorem 3.17, we have the following useful result.

**Corollary 4.23** (Scott [176]/Jockusch and Soare [105]). *There is an $\omega$-model of $\mathrm{WKL}_0$ consisting entirely of low sets.*

Since such a model does not contain $\emptyset'$, it is not a model of $\mathrm{ACA}_0$.

**Corollary 4.24** (Friedman [59]). $\mathrm{WKL}_0 \nvdash \mathrm{ACA}_0$.

Since KL is equivalent to $\mathrm{ACA}_0$ over $\mathrm{RCA}_0$, we see that König's Lemma is indeed strictly stronger than Weak König's Lemma over $\mathrm{RCA}_0$. It is worth reiterating that this fact gives us more information than the computability theoretic fact that Weak König's Lemma has low solutions, while König's Lemma does not, as it implies that, not only can we not in general obtain solutions for König's Lemma by single applications of Weak König's Lemma, we cannot do it even with multiple iterated applications of Weak König's Lemma. On the other hand, the computability theoretic fact that König's Lemma has $\Delta_3^0$ solutions gives us more information than the reverse mathematical fact that König's Lemma is provable in $\mathrm{ACA}_0$, which implies only that König's Lemma has arithmetic solutions. These examples illustrate the fact that computability theoretic and reverse mathematical results give related but often different information, and in many cases both are needed for a complete picture.

The hyperimmune-free basis and cone-avoidance basis theorems give us the following additional consequences of Theorem 4.22.

**Corollary 4.25** (Scott [176]/Jockusch and Soare [105]). *There is an $\omega$-model of $\mathrm{WKL}_0$ consisting entirely of sets of hyperimmune-free degree. For every noncomputable set $X$, there is an $\omega$-model of $\mathrm{WKL}_0$ none of whose elements compute $X$.*

Let $\mathcal{M}$ and $\mathcal{N}$ be $\omega$-models of WKL$_0$ consisting entirely of low sets and of sets of hyperimmune-free degree, respectively. Then $\mathcal{M} \cap \mathcal{N}$ contains only the computable sets, since every low set is $\Delta_2^0$ and no noncomputable set of hyperimmune-free degree can be $\Delta_2^0$ (as mentioned following Theorem 3.13). Since the computable sets do not form an $\omega$-model of WKL$_0$, it follows that there is no minimal $\omega$-model of WKL$_0$.

Recall the relation $\gg$ introduced in Section 3.1, where $A \gg B$ means that $A$ has PA degree relative to $B$. The relativized form of Theorem 4.22 says that if $A \gg B$ then there is an $\omega$-model of WKL$_0$ containing $B$ and consisting entirely of $A$-computable sets. This model contains a set $C \gg B$, and hence also a set $D \gg C$. Since $A \geqslant_T D$, we also have $A \gg C$. Thus we have the following result, which will be useful below.

**Theorem 4.26** (Simpson [185]). *If $A \gg B$ then there is a $C$ such that $A \gg C \gg B$.*

Another interesting consequence of the relativized form of Theorem 4.22 is the following. Let $P$ be a principle expressed by a sentence of the form $\forall X \, [\theta(X) \rightarrow \exists Y \, \psi(X,Y)]$, with $\theta$ and $\psi$ arithmetic, such that every $\omega$-model of WKL$_0$ is a model of $P$. Let $X$ be an instance of $P$ and let $T$ be an $X$-computable infinite binary tree such that every path on $T$ has PA degree over $X$. If $A$ is such a path, then there is an $\omega$-model of WKL$_0$, and hence of $P$, containing $X$ and consisting entirely of $A$-computable sets, so there is an $A$-computable solution to $X$. Thus $T$ witnesses the fact that $P \leqslant_c$ WKL.

There is a different general method for constructing $\omega$-models that yields Corollaries 4.23 and 4.25 without appealing to Theorem 4.22. Let $P$ be a true principle expressed by a sentence of the form $\forall X \, [\theta(X) \rightarrow \exists Y \, \psi(X,Y)]$, with $\theta$ and $\psi$ arithmetic (for instance, Weak König's Lemma). In analyzing the strength of $P$, we often begin by trying to find a small downwards-closed complexity class $\mathcal{C}$ (e.g., the class of computable sets, the class of low sets, the class of arithmetic sets) such that

$$\forall \text{ computable } X \, [\theta(X) \rightarrow \exists Y \in \mathcal{C} \, \psi(X,Y)]. \tag{4.4}$$

Informally, we say that "$P$ has solutions in $\mathcal{C}$." Thus, for example, the low basis theorem implies that Weak König's Lemma has low solutions.

It is usually the case that, for any oracle $X$, the class $\mathcal{C}$ can be relativized to a class $\mathcal{C}^X$ (e.g., the class of $X$-computable sets, the class of sets that are low relative to $X$, the class of sets that are arithmetic relative to $X$), and the proof of (4.4) can be relativized to show that

$$\forall Z \, \forall X \leqslant_T Z \, [\theta(X) \rightarrow \exists Y \, [Z \oplus Y \in \mathcal{C}^Z \wedge \psi(X,Y)]].$$

Such a result can be particularly useful if $\mathcal{C}$ is closed under relativization, in the sense that if $Z \in \mathcal{C}$ then $\mathcal{C}^Z = \mathcal{C}$. (For example, the class of low sets is closed under relativization, while the class of $\Delta_2^0$ sets is not.) In this case, we have

$$\forall Z \in \mathcal{C} \, \forall X \leqslant_{\mathrm{T}} Z \, [\theta(X) \to \exists Y \, [Z \oplus Y \in \mathcal{C} \wedge \psi(X, Y)]], \qquad (4.5)$$

and we can build an $\omega$-model of $\mathrm{RCA}_0 + P$ consisting entirely of sets in $\mathcal{C}$ as follows.

**Proposition 4.27.** *Let $\mathcal{C}$ be a nonempty collection of sets that is closed downwards under Turing reducibility, and let $P$ be a principle expressed by a sentence of the form $\forall X \, [\theta(X) \to \exists Y \, \psi(X, Y)]$, with $\theta$ and $\psi$ arithmetic. If (4.5) holds, then $\mathrm{RCA}_0 + P$ has an $\omega$-model consisting entirely of sets in $\mathcal{C}$.*

*Proof.* We proceed in stages. First, let $Z_0 = \emptyset$. Note that $Z_0 \in \mathcal{C}$. At the beginning of stage $s$, we are given a set $Z_s \in \mathcal{C}$. We will arrange the construction so that $Z_0 \leqslant_{\mathrm{T}} Z_1 \leqslant_{\mathrm{T}} \cdots$. Let $X_{\langle s,0 \rangle}, X_{\langle s,1 \rangle}, \ldots$ be a (noneffective) list of all $Z_s$-computable sets $X$ such that $\theta(X)$ holds.

Let $\langle t, n \rangle$ be the least pair such that $t \leqslant s$ and we have not yet acted for $X_{\langle t,n \rangle}$. Since $X_{\langle t,n \rangle} \leqslant_{\mathrm{T}} Z_t \leqslant_{\mathrm{T}} Z_s \in \mathcal{C}$, there is a $Y$ such that $Z_s \oplus Y \in \mathcal{C}$ and $\psi(X_{\langle t,n \rangle}, Y)$ holds. Let $Z_{s+1} = Z_s \oplus Y$ and say that we have acted for $X_{\langle t,n \rangle}$.

Finally, let $\mathcal{M}$ be the $\omega$-model consisting of all sets $X$ such that $X \leqslant_{\mathrm{T}} Z_s$ for some $s$. Notice that $\mathcal{M}$ consists entirely of sets in $\mathcal{C}$. Clearly, $\mathcal{M}$ is a Turing ideal, and hence is a model of $\mathrm{RCA}_0$. Let $X \in \mathcal{M}$ be such that $\mathcal{M} \vDash \theta(X)$. Since $\theta$ is arithmetic and $\mathcal{M}$ is an $\omega$-model, $\theta(X)$ holds. Then $X = X_{\langle t,n \rangle}$ for some $t$ and $n$. Thus, there is a stage $s$ at which we act for $X$. At this stage, we ensure that there is a $Y \in \mathcal{M}$ such that $\psi(X, Y)$ holds. Since $\psi$ is arithmetic and $\mathcal{M}$ is an $\omega$-model, $\mathcal{M} \vDash \psi(X, Y)$. Thus $\mathcal{M}$ is an $\omega$-model of $P$. $\qquad\square$

Since the low basis theorem holds in relativized form and the class of low sets is closed under relativization, we have an alternate proof of Corollary 4.23. Corollary 4.25 also follows in a similar manner.

Notice that the above proof involves a noneffective iteration of an effective property. The ability to argue noneffectively about $\omega$-models is often crucial. One useful consequence of this iterative construction of $\omega$-models is that it can handle multiple principles at once.

26 **Exercise 4.28.** Let $\mathcal{C}$ be a nonempty collection of sets that is closed downwards under Turing reducibility. For each $i \in \omega$, let $Q_i$ be a principle expressed by a sentence of the form $\forall X\,[\theta_i(X) \rightarrow \exists Y\,\psi_i(X,Y)]$, with $\theta_i$ and $\psi_i$ arithmetic. Show that if for all $i$ we have

$$\forall Z \in \mathcal{C}\,\forall X \leqslant_{\mathrm{T}} Z\,[\theta_i(X) \rightarrow \exists Y\,[Z \oplus Y \in \mathcal{C} \wedge \psi_i(X,Y)]]$$

then $\mathrm{RCA}_0 + Q_0 + Q_1 + \cdots$ has an $\omega$-model consisting entirely of sets in $\mathcal{C}$.

The role of PA degrees in the study of $\mathrm{WKL}_0$ is fulfilled by 1-random degrees for $\mathrm{WWKL}_0$ (see Downey and Hirschfeldt [40] for the definition of 1-randomness). Clearly $\mathrm{WKL}_0 \vdash \mathrm{WWKL}_0$; the following exercise requires some knowledge of the theory of algorithmic randomness.

27 **Exercise 4.29** (Yu and Simpson [211]). Use algorithmic randomness to build an $\omega$-model of $\mathrm{WWKL}_0$ that does not contain any sets of PA degree, which implies that $\mathrm{WWKL}_0 \nvdash \mathrm{WKL}_0$.

## 4.6    First order axioms

We will not say much about first order axioms until Section 6.8, but they do play an important role in reverse mathematics. Indeed, determining the first order (i.e., number theoretic) consequences of second order theorems has been a major line of research, and we will see examples of this work later in this book. Given our purposes, we will work over $\mathrm{RCA}_0$, but the implications and equivalences we discuss in this section can be established in weak first order systems. For details and much more on first order arithmetic, see Hájek and Pudlák [75].

Particularly prominent in reverse mathematics are two classes of first order axioms: induction axioms, which we have already encountered, and bounding axioms, of which the $\Sigma_2^0$ version was mentioned near the end of Section 4.1, and the $\Sigma_1^0$ version in Exercise 4.8.

**Definition 4.30.** Let $\mathrm{I}\Sigma_n^0$ be the principle of $\Sigma_n^0$-induction, which we recall is expressed by the axiom scheme consisting of

$$(\varphi(0) \wedge \forall n\,[\varphi(n) \rightarrow \varphi(n+1)]) \rightarrow \forall n\,\varphi(n)$$

for each $\Sigma_n^0$ formula $\varphi$. The principle $\mathrm{I}\Pi_n^0$ is defined analogously.

Let $\mathrm{B}\Sigma_n^0$ (for $\Sigma_n^0$-*bounding*) be the principle expressed by the axiom scheme consisting of

$$\forall n\,[\forall i < n\,\exists k\,\varphi(i,k) \rightarrow \exists b\,\forall i < n\,\exists k < b\,\varphi(i,k)]$$

for each $\Sigma_n^0$ formula $\varphi$. The principle $\mathrm{B}\Pi_n^0$ is defined analogously.

One difference between nonstandard first order models of $P_0^-$ and the standard model is the existence of *proper cuts*, i.e., proper initial segments closed under the successor relation. Induction axioms say that such cuts cannot be too easily definable. $I\Sigma_n^0$, for instance, says that there are no $\Sigma_n^0$-definable proper cuts. Another property of nonstandard models is the existence of cofinal maps with bounded domains (where a *cofinal* map is one whose range is unbounded). Bounding axioms say that such maps cannot be too easily definable. $B\Sigma_n^0$, for instance, implies that there is no $\Sigma_n^0$-definable function whose domain is a proper initial segment of the natural numbers, but whose range is unbounded in the natural numbers.

These restrictions on the "pathologies" of a nonstandard model have very useful consequences. For example, suppose we are working in RCA$_0$ + $B\Sigma_2^0$, and let $P$ be a $\Delta_2^0$ predicate, with an approximation $f$ given by the limit lemma. That is, for each $n$, we have that $\lim_s f(n, s)$ exists and $P(n)$ iff $\lim_s f(n, s) = 1$. Let $g(n)$ be the stage at which the approximation to $P(n)$ settles, i.e., the least $t$ such that $f(n, s) = f(n, t)$ for all $s \geqslant t$. Then $g(n)$ is a $\Sigma_2^0$-definable function, so by $B\Sigma_2^0$, for each $m$, the range of $g \restriction m$ is bounded. In other words, there is a $t$ such that $f(n, s) = f(n, t)$ for all $n < m$ and $s \geqslant t$. In particular, for each $n < m$, we have $P(n)$ iff $f(n, t) = 1$, so by $\Sigma_0^0$-comprehension, there is a set $F = \{n < m : P(n)\}$. Thus we see that, in the presence of $B\Sigma_2^0$, approximations to $\Delta_2^0$ predicates converge not only pointwise, but on initial segments, and we have *bounded $\Delta_2^0$-comprehension*, i.e., for each such predicate $P$ and each $m$, there is a (finite) set corresponding to $P \restriction m$. (Predicates with the latter property are called *amenable*.) It is not difficult to imagine how these properties can be quite useful in computability theoretic arguments. Without $B\Sigma_2^0$, it is possible to have both of them fail for some $\Delta_2^0$ predicate.

Note that $I\Sigma_n^0$ is clearly equivalent to the *least number principle for $\Pi_n^0$ formulas*, the axiom scheme stating that, for each $\Pi_n^0$ formula $\psi$, if $\psi(n)$ holds of some $n$, then there is a least $n$ such that $\psi(n)$ holds.

28 **Exercise 4.31** (Paris and Kirby [161]).
a. Show that $I\Sigma_n^0$ and $I\Pi_n^0$ are equivalent over RCA$_0$. [See Exercises 4.7 and 4.8.]
b. Show that $B\Sigma_n^0$ and $B\Pi_{n-1}^0$ are equivalent over RCA$_0$.
c. Show that $I\Sigma_n^0$ is equivalent over RCA$_0$ to bounded $\Sigma_n^0$-comprehension, the statement that for every $\Sigma_n^0$ predicate $P$ and every $m$, there is a set $A$ such that $n \in A$ iff $n < m \wedge P(n)$. [See Exercise 4.7.]

The principles in Definition 4.30 form a strict hierarchy. The following theorem was originally proved in the first order context, but because the first order part of $RCA_0$ is $\Sigma_1^0$-PA, it remains true when working over $RCA_0$ (or even $WKL_0$).

**Theorem 4.32** (Paris and Kirby [161]). *Over $RCA_0$ (or $WKL_0$), we have the following implications, all of which are strict:*

$$I\Sigma_1^0 \leftarrow B\Sigma_2^0 \leftarrow I\Sigma_2^0 \leftarrow B\Sigma_3^0 \leftarrow I\Sigma_3^0 \leftarrow \cdots . \qquad (4.6)$$

*In particular, none of these principles other than $I\Sigma_1^0$ is provable in $RCA_0$, or even $WKL_0$.*

$\boxed{29}$ **Exercise 4.33.** Prove the implications in Theorem 4.32.

For a proof of the nonimplications in Theorem 4.32, see Hájek and Pudlák [75]. As noted in Section 4.3, arithmetic induction holds in $ACA_0$, and hence so do all of the principles in (4.6).

As it turns out, the above bounding principles may also be seen as induction principles, due to the following theorem, which again was originally proved in the first order context, and helps explain the hierarchy (4.6).

**Theorem 4.34** (Slaman [195]). $B\Sigma_n^0$ *is equivalent over $RCA_0$ to $I\Delta_n^0$, the $\Delta_n^0$-induction scheme given in Definition 4.3.*

There are many other interesting first order principles of a similar nature to induction and bounding; [75] is a good source. Some of these principles turn out to be useful equivalents of induction and bounding principles. For instance, $B\Sigma_n^0$ is equivalent over $RCA_0$ to the *finite axiom of choice for $\Pi_n^0$ properties* (see [75]): for any sequence $n_0, \ldots, n_k$ and any $\Pi_n^0$ property $P$, if for each $i \leqslant k$ there is an $m$ with $P(n_i, m)$, then there is a sequence $m_0, \ldots, m_k$ such that $P(n_i, m_i)$ for each $i \leqslant k$. Similarly, $I\Sigma_n^0$ is equivalent over $RCA_0$ to the *finite $\Pi_n^0$ recursion principle* (see Hirschfeldt and Shore [88], where this principle is called $PREC_n$): if $P$ is a $\Pi_n^0$ property defining a total function, then for each $z$ and $k$, there is a sequence $x_0, \ldots, x_k$ such that $x_0 = z$ and $P(x_i, x_{i+1})$ holds for all $i < k$. (We have discussed the $n = 1$ case in Section 4.1.)

In addition, these first order principles have many combinatorial equivalents. The following example will be useful in Section 9.3. Let $T$ be a binary tree with no dead ends. A node $\sigma \in T$ is an *atom* if for each $n > |\sigma|$, there is exactly one extension of $\sigma$ of length $n$ in $T$. The tree $T$ is *atomic* if each node of $T$ can be extended to an atom, and *strongly atomic* if for every $\sigma_0, \ldots, \sigma_n \in T$, each $\sigma_i$ can be extended to an atom.

**30** **Exercise 4.35** (Hirschfeldt, Lange, and Shore [87]). Show that $B\Sigma_2^0$ is equivalent over $RCA_0$ to the statement that every atomic tree with no dead ends is strongly atomic.

Note that, despite being equivalent to $B\Sigma_2^0$, the statement in the above exercise is a second order statement. It is also possible to have second order statements that live in the "first order part" of the reverse mathematical universe, yet are not equivalent to any first order statement. For instance, Hirschfeldt, Lange, and Shore [87] gave an example of a statement, called $\Pi_1^0 GA$, that is provable in $RCA_0 + I\Sigma_2^0$ but not in $RCA_0$, and does not imply any first order statements not provable in $RCA_0$. (This principle is closely related to the principle $\Pi_1^0 G$ in Definition 9.44 below.)

## 4.7 Further remarks

We have already referred to the "big five" systems of reverse mathematics: $RCA_0$, $WKL_0$, $ACA_0$, $ATR_0$, and $\Pi_1^1\text{-}CA_0$. Many examples of theorems in various areas of mathematics equivalent to these systems can be found in Simpson [191]. The picture presented by such equivalences is particularly neat because these systems are linearly ordered by strength (and this fact remains true if we add $WWKL_0$ to the list). As we will see below, though, there are many other theorems that do not fit this picture, even if we set aside issues arising from the need for additional first order principles in some cases. (I will have more to say about this situation in the following chapter.) I believe it is still an open question exactly how close to the truth this neat view of reverse mathematics will seem in the future, as we investigate more and more areas of mathematics, and increasingly develop methods for discovering and analyzing theorems that fall outside the big five picture. Nevertheless, I do think we have enough evidence that this phenomenon of a small number of systems, linearly ordered by strength, that capture a large swath of mathematics is undoubtedly a real and interesting one, especially given the connections between these systems and major foundational programs, as discussed for instance in Simpson [190, 191]. As Montalbán [146] puts it, "Though we have some sense of why this phenomenon occurs, we really do not have a clear explanation for it, let alone a strictly logical or mathematical reason for it. The way I view it, gaining a greater understanding of this phenomenon is currently one of the driving questions behind reverse mathematics."

Montalbán [146] also mentions two noteworthy related issues. One is the distinction between "robust" and "non-robust" systems. The idea there is that the big five systems (and $WWKL_0$) share the property that small variations in the way they or their equivalents are stated tend to yield statements that remain equivalent to the original versions. As we will see below, this is not the case for many other systems. The search for new robust systems is an important research program in reverse mathematics. (See the discussion following Theorem 10.23 below.) Another issue mentioned in [146] is the relative consistency strength of systems of reverse mathematics, and the related concept of interpretability between systems. Considering these measures, rather than just relative strength, leads to a tamer picture of the reverse mathematical universe, even when principles like the ones we will discuss below are brought into consideration. See [146] and Simpson [190,191] for more on this issue. Here the proof theoretic perspective of ordinal analysis becomes particularly important; see for instance Rathjen [167].

There are also theorems that can be stated in the language of second order arithmetic and proved in ZFC, but cannot be proved even in full second order arithmetic. For a recent example of a family of (determinacy) principles that stretches the limits of second order arithmetic and eventually escapes its bounds entirely, see Montalbán and Shore [148]. See also the discussion in Shore [183].

Not every result of computable mathematics has a reverse mathematical analog. For example, recall from Exercise 3.28 that every path on a computable binary tree with finitely many paths is computable. There does not seem to be a reasonable way to obtain an analog to this statement in the context of reverse mathematics. On the other hand, there is sometimes more than one reasonable way to formalize a theorem, and different formalizations can have different strengths. This phenomenon is a feature, not a bug. Highlighting differences between statements that may seem interchangeable in ordinary mathematical practice is one of the important applications of our metamathematical approach. An example is the difference between König's Lemma and the statement in Exercise 4.14, which might seem entirely equivalent but is in fact at the level of $WKL_0$, while KL is at the level of $ACA_0$. This nonequivalence points out the important computational difference between knowing something is bounded and actually having access to a bound. Another example discussed above is given by the versions of Lindenbaum's Lemma (or Gödel's Completeness Theorem) for sets of sentences, which is equivalent to $WKL_0$, and for theories, which is provable in $RCA_0$. Yet another example, which we will discuss in Section 9.3, is the difference between the statements "Every complete atomic theory in a countable language has a countable atomic model." and "Every complete atomic theory in a countable language has a countable prime model." There the issue is that the equivalence between being prime and being atomic for countable models is not provable in $RCA_0$, although sometimes, in the countable setting, one tends to use the terms interchangeably.

Of course, not all possible formalizations are equally interesting. For example, in formalizing (W)KL, we chose to define a tree as a subset of $\mathbb{N}^{<\mathbb{N}}$ (or $2^{<\mathbb{N}}$) that is closed under prefixes. One could also (and often does) define a tree as a connected acyclic graph with a distinguished root node. In the binary case, there is a major computational difference between these two definitions: for a computable binary tree in our original definition, we can always compute how many immediate successors a given node has, but that is not necessarily the case with a computable binary tree in the graph sense. Thus it is not difficult to build a computable infinite binary tree in the graph sense all of whose paths compute $\emptyset'$. So we see that taking the graph definition of a tree elides the important computability theoretic and reverse mathematical distinction between WKL and KL. The wealth of theorems in various areas of mathematics that are at the $WKL_0$ level gives further evidence that, in this context, our choice of formalization for (W)KL is the preferred one.

Another issue worth considering is the strength of different proofs of the same theorem. We have already seen that the most obvious proof of a theorem is not necessarily the most effective one. For instance, we can prove WKL using the simple argument given at the beginning of Section 3.2, but cannot get either the existence of low solutions to WKL or the fact that $WKL_0$ does not imply $ACA_0$ from that argument. In this case, we actually have a rigorous proof of this fact: the same proof is also a proof of KL, and KL does not have low solutions in general, and does imply $ACA_0$ over $RCA_0$. But in general, it is a more difficult and subtle matter to calibrate the strength of a proof, as opposed to that of the theorem it proves. Of course, if one has a completely formal proof, one can find out exactly what axioms it employs, but for informal proofs the situation is often muddier. Consider a proof by induction in which we show that every $n$ has some $\Sigma_1^0$ property $P$. We could form the set of all $n$ with property $P$ and proceed by set induction, or not form the set and proceed by $\Sigma_1^0$-induction. Generally, we would think of these as "the same proof", but the former version cannot be carried out in $RCA_0$, while the latter can.

Of course, as we have discussed above, theorems can also have multiple versions that are "classically equivalent" but different from the points of view of computability theory and reverse mathematics. But in most cases there are only a few reasonable versions (and often even only one clearly most natural one), and we can list out and examine all of them. Proofs are much longer and more complicated objects, and in many cases all we can say about a given proof is that it does not look like it can be carried out in a given system $S$, while not being able to rule out in any rigorous sense that a complicated series of manipulations like the one mentioned above, where we replace $\Sigma_1^0$-comprehension plus set induction by $\Sigma_1^0$-induction, might yield a proof in $S$ that we would consider essentially the same as the original one. Of course, as we have seen, an obvious exception is when we can show that the proof yields a theorem that we can show is not provable in $S$ (for instance by building an appropriate $\omega$-model). We will see a good example of multiple proofs of the same theorem with quite different computability theoretic and reverse mathematical strengths in Section 6.1.

# Chapter 5

# In Defense of Disarray

Aesthetically, I am allergic to neatness. This preference for a certain amount of disarray is in line with (or perhaps engenders...) my epistemological views. I tend to believe that, in any interesting setting, when the truth has been fully systematized and organized, the truth has long since lain elsewhere. I do believe in organization and classification, but my systems are ones in which the organizing principles themselves are complicated enough to need organization, where connections range all the way from the completely tight to the very tenuous, and where the great engine of ineffability is continually generating.

Much has been made of the ubiquity of the "big five" systems of reverse mathematics. It is indeed the case that an exceptional number of theorems that have been investigated in reverse mathematics have turned out to be equivalent to one of these systems, which, as mentioned above, is without a doubt a fascinating phenomenon, and worthy of attention. Much of this book, however, will give a different picture of the reverse mathematical universe, one that is much less neat, and can seem rather chaotic and devoid of structure. We will examine several principles that are not equivalent to any of the big five systems, and have complicated interrelationships.

Even in the world of the big five systems, the computability theoretic perspective can reveal further complexity. Equivalence to $ACA_0$, in particular, is a concept that elides many interesting distinctions. For instance, a principle whose instances require one jump to solve and one whose instances require several jumps to solve will both end up at the level of $ACA_0$. We will see a particular example of this phenomenon in Section 6.2, where we will study Ramsey's Theorem for colorings of unordered $n$-tuples of natural numbers. We will see that, for each $n \geqslant 3$, the corresponding principle is equivalent to $ACA_0$, but that the complexity of the problem of finding solutions to instances of these principles goes up with $n$, in terms of the arithmetic hierarchy, reflecting the induction on $n$ involved in their proof.

As one might imagine given the first paragraph of this chapter, I see this complexity as a positive feature of reverse mathematics and computable mathematics. But one should never mistake lack of obvious structure for actual lack of structure. I also believe there is something holding together and structuring the results we will study below, though it is not as clean or easily graspable as the big five systems picture of the universe. To give an idea of what I think this something is, and whence it arises, let us take a detour through what may seem like a somewhat abstruse question.

What is computability theory about? There is likely no single answer that will satisfy everyone working in the field, and many might not even consider questions at this level of generality and abstraction to be interesting, but let us consider a couple of possible answers anyway.

The most obvious one is that computability theory is about computation. But computability theorists routinely ignore many clearly significant aspects of the notion of computation, such as time and space bounds. It strikes me as difficult to argue that a field that in the vast majority of cases considers all computable sets to be the same is studying (rather than merely using) the notion of computation.

Another possible answer to our question is that computability theory is about definability. I first saw this viewpoint explicitly espoused in Slaman's Gödel Lecture at the 2001 Annual Meeting of the Association for Symbolic Logic. In the slides for that lecture [194], he defined computability theory (or to be true to his terminology, recursion theory) as "[t]hat part of mathematical logic which is focused on definability, especially for subsets of the natural numbers ($\omega$) and of the real numbers ($2^\omega$)." I think the subject of this book bears witness to the reasonableness of this definition.

To me, however, an even more satisfying answer to our question is that computability theory is about the relationship between computation and definability. Each of these notions is extremely natural and fundamental on its own, but to my mind, it is the fact that they come together so precisely, as embodied in results such as Post's Theorem 2.3 and Exercise 4.19, that gives computability theory its life and its identity. And from this relationship comes what I think is a valuable heuristic: Computability theoretically natural notions tend to be combinatorially natural.

Iterates of the jump form an obvious example. The halting problem relative to the halting problem might seem like an esoteric object, but $\emptyset''$ is a $\Sigma_2^0$-complete set (i.e., it is $\Sigma_2^0$ and can compute any $\Sigma_2^0$ set), and hence embodies exactly the computational power needed to answer natural $\Sigma_2^0$ questions such as whether a given computable (or c.e.) set is infinite. So whenever issues of finite vs. infinite arise, the double jump is likely at least hovering in the background, and is often front and center. But suppose that what the combinatorics of a problem requires is not determining whether a given set is infinite, but, given two sets, at least one of which is infinite, choosing one that is infinite. Then the notion that comes to the fore is that of PA degrees relative to $\emptyset'$, a concept that might at first seem even stranger than the double jump. We will see an example of exactly this phenomenon in Section 6.5, in connection with the combinatorics of building cohesive sets (see Definition 6.30).

Or consider the notion of lowness. Its definition may make it seem rather technical, but notice that a set $X$ is low iff every set that is $\Sigma_2^0$ relative to $X$ is $\Sigma_2^0$ (and similarly for $\Pi_2^0$ and $\Delta_2^0$). In other words, the low sets are the ones that do not contain enough information to give us any new definitions at the level of two (first order) quantifiers. As already mentioned, the two quantifier level is the level of central mathematical concepts such as the notion of infinity, so it is not surprising then that lowness arises naturally in the study of many mathematical principles and constructions. In similar ways, notions such as low₂ness, hyperimmunity, and so on make themselves invaluable in metamathematical analysis.

So it is these notions and their relationships, expressing the interplay between computability and definability, that provide the structure for the growing collection of results in reverse mathematics and computable mathematics of which we will give examples below. Of course, these relationships are themselves complex, so it is not necessarily easy to see the patterns at first, or to describe them in general terms. But it is fascinating to see them emerge as one immerses oneself in the area. And, fortunately, we do have tools for organizing the organizing principles themselves, tools such as the structure of the Turing degrees.

There is a criticism that has sometimes been leveled at the study of the degrees: All natural individual mathematical objects whose degrees are known live in a very small class of degrees. For arithmetic objects, these are the degrees of the iterates of the jump. Indeed, I imagine most computability theorists (myself included) would agree that it is most likely the case that one will never find a natural arithmetic set in any other degree. The word "natural" is a contentious one, of course, but one can take as a working definition something like "a set is natural if it is of interest to mathematicians independently of its metamathematical (in particular, computability theoretic) properties." So examples of natural sets in the same degree as $\emptyset'$ include the set of all Diophantine equations with integer solutions (the famous DPRM Theorem, giving a negative solution to Hilbert's 10th Problem; see Poonen [162]) and the set encoding the word problem for finitely presented groups (as shown by Novikov [156] and Boone [8]). We can take a given c.e. set $A$ and encode it into a finitely presented group $G$ so that the word problem in $G$ has the same degree as $A$, but neither $G$ nor its particular word problem would likely be of interest for any independent reason.

Without "real-world" examples of sets at any but a small number of very well-behaved degrees, why spend as much time studying the structure of the Turing degrees as computability theorists have? I believe that to answer this question, we need to shift focus from individual degrees to classes of degrees, such as the low degrees, the PA degrees, the hyperimmune-free degrees, and so on. These classes (and even "weirder" ones, such as the class of sets whose jumps have PA degree relative to $\emptyset'$) do often correspond to natural classes of mathematical objects (e.g., the PA degrees and solutions to instances of principles at the level of $\mathrm{WKL_0}$). But because these classes interact in complex ways, we need fine structure tools to analyze them, and the study of the degrees provides many such tools. Certainly, other structures that aim to capture natural classes of sets as individual objects, such as the Medvedev [135] and Muchnik [151] degrees of mass problems discussed for instance in [192], also offer useful perspectives. But because they ignore the fine structure of these classes with respect to the fundamental notion of computability, they should not be thought of as potential replacements for the study of the Turing degrees in computable mathematics, reverse mathematics, and other areas of application of computability theory.

Let us look at just one example, in the context of applications of computability theory to reverse mathematics. We will look at this particular example in greater detail in Section 9.3, and will also see other examples below. We have briefly discussed the hyperimmune-free degrees in Section 3.1. A degree that is not hyperimmune-free is *hyperimmune*. We have mentioned the fact that all noncomputable $\Delta_2^0$ sets have hyperimmune degree. This fact relativizes, so in particular, if $A$ and $B$ are c.e. sets such that $A <_\mathrm{T} B$, then $B$ has hyperimmune degree relative to $A$. Let us now bring up one of the classic results of the study of the c.e. degrees (i.e., the degrees of c.e. sets), the Sacks Density Theorem.

Early on in the development of computability theory, there were only two c.e. degrees known, the degree of the computable sets, and the degree of the halting problem. Post [164] asked whether there are any others. Post's Problem was answered independently by Friedberg [57] and Muchnik [150]. The method they introduced in their proofs, known as the priority method, became one of the hallmarks of computability theory, much like forcing in set theory following the work of Cohen (see Kunen [118]). (We will give an example of a priority argument in computable mathematics in the proof of Theorem 9.11 below.) Using this method, computability theorists began to reveal the complex structure of the c.e. degrees. In particular, Sacks [174] showed that the c.e. degrees are dense. That is, if $A <_T B$ are c.e. sets, then there is a c.e. set $C$ such that $A <_T C <_T B$.

Post's Problem was clearly a natural one, given that c.e. sets permeate mathematics, but, absent any natural examples of incomplete noncomputable c.e. sets, one might wonder what information the Sacks Density Theorem really gives us. One of the many answers to this question is that it tells us that for any noncomputable c.e. set $X$, there is an $\omega$-model $S$ of RCA$_0$ consisting entirely of $X$-computable sets, and such that if $A \in S$ then there is a $B \in S$ that has hyperimmune degree relative to $A$. This model is built as in Proposition 4.27: Start with $Z_0 = \emptyset$. Having defined the c.e. set $Z_n <_T X$, let $Z_{n+1}$ be a c.e. set such that $Z_n <_T Z_{n+1} <_T X$, which exists by the Sacks Density Theorem. Then $Z_{n+1}$ has hyperimmune degree relative to $Z_n$ (and hence relative to any $Z_n$-computable set). Now let $S = \{Y : \exists n\, Y \leqslant_T Z_n\}$.

So far, everything seems quite internal to degree theoretic concerns. But recall our heuristic. Hyperimmunity is a computability theoretic natural concept, so we should not be surprised to find that it can be characterized by other means. Indeed, Hirschfeldt, Shore, and Slaman [89] showed that a version of the Omitting Types Theorem of basic model theory, called OPT, is equivalent over RCA$_0$ to the statement that for every set $Y$, there is a set that has hyperimmune degree relative to $Y$ (see Section 9.3 for more details). Thus $S$ is an $\omega$-model of OPT. Since the set $X$ bounding $S$ is an arbitrary noncomputable c.e. set, and hence can be taken to be quite computability theoretically weak, this fact can be used to prove a wide range of nonimplication results between OPT and other principles, including WKL$_0$, since there are c.e. sets that do not compute any PA degree. (Conversely, WKL$_0$ also does not prove OPT, since the hyperimmune-free basis theorem allows us to build a model of WKL$_0$ consisting entirely of sets of hyperimmune-free degree.)

Of course, I find structural results such as the Density Theorem of intrinsic interest. But the point of the above example, and many others like it, is that such results are also of great importance to those whose interest in notions such as computability and computable enumerability (or their alter egos, $\Delta^0_1$-definability and $\Sigma^0_1$-definability) is purely due to their applications to metamathematical analysis.

But enough of this high-level discussion. If one is to understand the above concerns more deeply, and reach an informed opinion on the merits of my arguments, fully worked out examples are needed. So, with all of this in mind, let us proceed.

# Chapter 6

# Achieving Consensus: Ramsey's Theorem

Ramsey's Theorem [166] captures the idea that total disorder is impossible: a sufficiently large structure will always contain a large ordered substructure. A well-known special case of the finite version is that in any group of six people, there are at least three people who all know each other or three people who are all strangers to each other. We will examine the following infinite versions of Ramsey's Theorem. As we will see, the computability theoretic and reverse mathematical analysis of these and related principles makes for an excellent case study, as it involves a range of central techniques and ideas in these areas.

**Definition 6.1.** For a set $X$, let $[X]^n$ be the collection of $n$-element subsets of $X$. A *k-coloring* of $[X]^n$ is a map $c : [X]^n \to k$. A set $H \subseteq X$ is *homogeneous for $c$* if there is an $i < k$ such that $c(s) = i$ for all $s \in [H]^n$. We also say that $H$ is *homogeneous to $i$*.

1. *Ramsey's Theorem for n-tuples and k colors* ($\mathrm{RT}^n_k$) is the statement that every $k$-coloring of $[\mathbb{N}]^n$ has an infinite homogeneous set.
2. *Ramsey's Theorem for n-tuples* ($\mathrm{RT}^n_{<\infty}$) is the statement $\forall k \geqslant 1\, \mathrm{RT}^n_k$.
3. *Ramsey's Theorem* (RT) is the statement $\forall n \geqslant 1\, \mathrm{RT}^n_{<\infty}$.

Note that the tuples in Ramsey's Theorem are unordered. Nevertheless, for simplicity of notation, we write $c(x_0, \ldots, x_n)$ for $c(\{x_0, \ldots, x_n\})$. Note also that, although $\mathrm{RT}^n_k$ is stated in terms of colorings of $[\mathbb{N}]^n$, it is clearly equivalent (over $\mathrm{RCA}_0$) to the statement that, for any infinite $X \subseteq \mathbb{N}$, every $k$-coloring of $[X]^n$ has an infinite homogeneous set.

We begin by noting that there are certain easy implications between the principles in Definition 6.1. In addition to the obvious ones, we have the following.

$\boxed{31}$ **Exercise 6.2.** Show that for every $n \geqslant 1$ and $k \geqslant 2$, the following are provable in $\mathrm{RCA}_0$.

**a.** $\mathrm{RT}^1_k$. [See Section 4.1.]
**b.** $\mathrm{RT}^{n+1}_k \to \mathrm{RT}^n_k$ and $\mathrm{RT}^{n+1}_{<\infty} \to \mathrm{RT}^n_{<\infty}$.
**c.** $\mathrm{RT}^n_k \to \mathrm{RT}^n_{k+1}$.

Note that item c in the above exercise implies that for all $k \in \omega$, we have

$$\mathrm{RCA}_0 \vdash \mathrm{RT}^n_2 \to \mathrm{RT}^n_k, \tag{6.1}$$

but it does not immediately imply that

$$\mathrm{RCA}_0 \vdash \mathrm{RT}^n_2 \to \mathrm{RT}^n_{<\infty}. \tag{6.2}$$

While (6.1) is proved by induction on $k$, the induction is not itself carried out within $\mathrm{RCA}_0$, which is what would be needed to prove (6.2). We will see that (6.2) does hold for $n \geqslant 3$, but does not hold for $n = 1$ or $n = 2$.

Since the $k = 1$ case is trivial, the above exercise shows that, from the perspective of reverse mathematics, we have only the following cases to consider: $\mathrm{RT}^n_2$ for $n \geqslant 2$, $\mathrm{RT}^n_{<\infty}$ for $n \geqslant 1$, and RT. The computability theoretic picture is less clear. Recall the notion of computability theoretic reduction $\leqslant_\mathrm{c}$ from Section 2.2. Clearly, for any $n$ and $j < k$, we have $\mathrm{RT}^n_j \leqslant_\mathrm{c} \mathrm{RT}^n_k$, and each $\mathrm{RT}^1_k$ is computably true, but the following question is open.

▶ **Open Question 6.3.** For which $n > 1$ and $1 < j < k$ is it the case that $\mathrm{RT}^n_k \leqslant_\mathrm{c} \mathrm{RT}^n_j$?

The usual solution to the last part of Exercise 6.2 does not help here, because in proving, for instance, that $\mathrm{RT}^n_2$ implies $\mathrm{RT}^n_3$, it uses two applications of $\mathrm{RT}^n_2$. For the notion of uniform computability theoretic reducibility $\leqslant_\mathrm{u}$ from Section 2.2, Hirschfeldt and Jockusch [85] have shown that if $n \geqslant 1$ and $1 < j < k$ then $\mathrm{RT}^n_k \not\leqslant_\mathrm{u} \mathrm{RT}^n_j$, extending a result of Dorais, Dzhafarov, Hirst, Mileti, and Shafer [38], who proved the analogous fact for the notion of strong uniform reducibility mentioned in Section 2.2.

## 6.1 Three proofs of Ramsey's Theorem

When one begins to study the strength of a mathematical theorem or family of theorems, usually the natural starting point is to look at the classical proof (or proofs). We will give three proofs of Ramsey's Theorem. Although they are all closely related, they have different levels of suitability for our purposes. The first and perhaps simplest is the original proof of the theorem, but as we will see, does not seem amenable to effectivization. The second yields a proof of $\mathrm{RT}^n_{<\infty}$ in $\mathrm{ACA}_0$. The third also yields such a proof, and in a more easily verifiable way. In addition, as we will see in the following sections, it can also be easily effectivized in a way that has several computability theoretic consequences.

*Proof of Ramsey's Theorem, version 1.* To prove $\mathrm{RT}^n_k$, we fix $k$ and proceed by induction on $n$. The $\mathrm{RT}^1_k$ case is trivial. Now suppose that $\mathrm{RT}^{n-1}_k$ holds and let $c : [\mathbb{N}]^n \to k$. The idea is first to build an infinite set $A$ such that for all $s \in [A]^n$, the color of $s$ depends only on the least element of $s$, and then to thin out $A$ into an infinite homogeneous set.

We define the elements $a_0 < a_1 < \cdots$ of $A$ by recursion. Let $a_0 = 0$. Let $d_0 : [\mathbb{N} \setminus \{a_0\}]^{n-1} \to k$ be defined by $d_0(s) = c(s \cup \{a_0\})$ and let $H_0$ be an infinite homogeneous set for $d_0$ such that $a_0 < \min H_0$. Let $c_0$ be the color to which $H_0$ is homogeneous.

Now let $a_1$ be the least element of $H_0$. Let $d_1 : [H_0 \setminus \{a_1\}]^{n-1} \to k$ be defined by $d_1(s) = c(s \cup \{a_1\})$ and let $H_1$ be an infinite homogeneous set for $d_1$ such that $a_1 < \min H_1$. Let $c_1$ be the color to which $H_1$ is homogeneous, and let $a_2$ be the least element of $H_1$.

Continuing in this way, we define $A = \{a_0 < a_1 < \cdots\}$. If $s \in [A]^n$ then let $i$ be least such that $a_i \in s$. Then all other elements of $s$ are in $H_i$, so $c(s) = c_i$. There is a $j < k$ such that $c_i = j$ for infinitely many $i$. Let $H = \{a_i : c_i = j\}$. Then $c(s) = j$ for all $s \in [H]^n$, so $H$ is an infinite homogeneous set for $c$. $\qquad\square$

While the above proof is conceptually simple, it is computationally complex. The real issue is the definition of $A$, since we can compute $H$ from $A$. For the $n = 2$ case, each $H_i$ can be taken to be computable, and $A$ can still be arithmetic:

**32** **Exercise 6.4.** Show that if $n = 2$ and $c$ is computable, then the set $A$ in the above proof can be $\emptyset''$-computable.

But let us now consider the general case. To define $H_0$, we need to appeal to the $n - 1$ case, so $H_0$ will in general not be computable. Indeed we will see in Theorem 6.13 that there are computable colorings of triples all of whose infinite homogeneous sets compute $\emptyset'$. Now $d_1$ is no longer a computable coloring (since its domain is $H_0 \setminus \{a_0\}$), but merely $H_0$-computable. Thus $H_1$ is obtained from the $n$ case *relativized to $H_0$*, whence $H_1$ might be more complex than $H_0$ (say at the $\emptyset''$ rather than the $\emptyset'$ level). As we continue the recursion, the sets $H_i$ might get more and complex, even if we assume that the $n-1$ case can be done arithmetically, reaching all the way up the arithmetic hierarchy. Since $A$ requires all the $H_i$ for its definition, we have little hope that $A$ will be arithmetic. Of course, the $n+1$ case will then be even worse, in that even $H_0$ might not be arithmetic.

In particular, it does not appear that the above proof can be carried out in $\mathrm{ACA}_0$. However, as mentioned in Section 4.7, an important theme in reverse mathematics is that different proofs of the same theorem can have vastly different computational content and reverse mathematical strength. Indeed, $\mathrm{RT}^n_{<\infty}$ *is* provable in $\mathrm{ACA}_0$ for each $n \geqslant 1$. (We will discuss the strength of the full theorem RT in Section 6.3.) One way to obtain a proof of this fact is to "turn around" the idea of our first proof of Ramsey's Theorem. Instead of constructing a set such that the color of any $n$-element subset depends only on its least element, which then allows us to use $\mathrm{RT}^1_k$ to find a homogeneous set, we construct a set such that the color of any $n$-element subset depends only on its least $n-1$ many elements, and then use $\mathrm{RT}^{n-1}_k$ to obtain our homogeneous set. The advantage of this approach is that instead of multiple applications of the $n-1$ case followed by a single application of the $n=1$ case, we now have multiple applications of the $n=1$ case followed by a single application of the $n-1$ case, resulting in a huge decrease in complexity. Thus we are led to the following notions.

**Definition 6.5.** Let $n \geqslant 2$. A set $A$ is *prehomogeneous* for $c : [\mathbb{N}]^n \to k$ if for all $s \in [A]^{n-1}$ and all $x, y \in A$ such that $x, y > \max s$, we have $c(s \cup \{x\}) = c(s \cup \{y\})$. In other words, a prehomogeneous set is one for which the color of an $n$ element subset depends only on the least $n-1$ many elements.

For a prehomogeneous set $A$, the coloring $d : [A]^{n-1} \to k$ *induced* by $c$ is defined by letting $d(s) = c(s \cup \{x\})$ for $s \in [A]^{n-1}$ and $x \in A$ such that $x > \max s$.

In the above definition, any set that is homogeneous for $d$ is also homogeneous for $c$, so we have the following fact, which is clearly provable in $\mathrm{RCA}_0$.

**Lemma 6.6.** *Let $n \geqslant 2$ and let $c : [\mathbb{N}]^n \to k$. If $\mathrm{RT}^{n-1}_k$ holds and $c$ has an infinite prehomogeneous set, then $c$ has an infinite homogeneous set.*

Thus, to prove Ramsey's Theorem, it is enough to show that for each $n \geqslant 2$, every $k$-coloring of $[\mathbb{N}]^n$ has an infinite prehomogeneous set. Then $\mathrm{RT}^n_{<\infty}$ follows by induction on $n$ (the base case $n = 1$ being trivial).

*Proof of Ramsey's Theorem, version 2.* Fix $n, k \geqslant 2$ and $c : [\mathbb{N}]^n \to k$. We show that $c$ has an infinite prehomogeneous set. We will define infinite sets $I_0 \supset I_1 \supset \cdots$, such that, letting $a_m = \min I_m$, we have $a_0 < a_1 < \cdots$, and if $s \in [\{a_i : i \leqslant m\}]^{n-1}$ and $x, y \in I_{m+1}$ then $c(s \cup \{x\}) = c(s \cup \{y\})$. It is then easy to see that $\{a_m : m \in \mathbb{N}\}$ is an infinite prehomogeneous set for $c$.

To build our sequence, first, for each $i < n-2$, let $I_i = \mathbb{N} \setminus [0, i]$. Now let $m \geqslant n - 3$ and suppose that we have defined $I_i$ (and hence $a_i = \min I_i$) for all $i \leqslant m$. Let $F = \{s \in [\{a_i : i \leqslant m\}]^{n-1}\}$. Let $d : I_m \setminus \{a_m\} \to k^F$ (where $k^F$ is the set of functions from $F$ into $k$) be defined by letting $d(x)$ be the function $s \mapsto c(s \cup \{x\})$. Thinking of $d$ as a $k^{|F|}$-coloring of $I_m \setminus \{a_m\}$, let $I_{m+1}$ be an infinite homogeneous set for $d$. It is now easy to check that our sequence has the desired properties. $\qquad\square$

In the above proof, each $I_m$ can be taken to be computable, but to pass from $I_m$ to $I_{m+1}$ requires us to know an element of $k^F$ whose preimage under $d$ is infinite. This is a two-quantifier question, which $\emptyset''$ can answer, so we can obtain a prehomogeneous set for a computable coloring $\emptyset''$-computably. Using this fact, we can obtain the following upper bound on the computability theoretic complexity of $\mathrm{RT}^n_{<\infty}$, which will be significantly improved in the following section.

33 **Exercise 6.7** (Manaster, see Jockusch [97]). Show that for every $n, k \geqslant 2$, every $k$-coloring of $[\mathbb{N}]^n$ has an $\emptyset^{(2n-2)}$-computable infinite homogeneous set.

With some more care, we can also use this idea to obtain a proof of $\mathrm{RT}^n_{<\infty}$ in $\mathrm{ACA}_0$. We will see another proof of this fact below.

34 **Exercise 6.8** (Simpson [188]). Adapt the above proof to work in $\mathrm{ACA}_0$, and use this fact to show that for each $n \in \omega$, we have $\mathrm{ACA}_0 \vdash \mathrm{RT}^n_{<\infty}$. [Assume the fact that $\mathrm{ACA}_0 \vdash \mathrm{RT}^1_{<\infty}$, which follows from the easy direction of Theorem 6.81 below.]

Note that we do not get a proof of the full RT in $\mathrm{ACA}_0$ because the induction on $n$ used to prove $\mathrm{RT}^n_{<\infty}$ in $\mathrm{ACA}_0$ is *external*, i.e., performed outside the system, and hence yields a separate proof in $\mathrm{ACA}_0$ for each (standard) $n$, not a single proof in $\mathrm{ACA}_0$ that works for all $n$. We do get a proof of RT in $\mathrm{ACA}_0$ together with $\Pi_2^1$-induction, though, since $\mathrm{RT}^n_{<\infty}$ is a $\Pi_2^1$ statement. In particular, RT is true in the minimal $\omega$-model of $\mathrm{ACA}_0$, and is arithmetically true. In the following sections, we will pinpoint the levels of the arithmetic hierarchy corresponding to our various versions of Ramsey's Theorem.

To attempt to further clarify the computability theoretic complexity of our second proof of Ramsey's Theorem, we can replace the process of choosing the infinite homogeneous sets $I_{m+1}$ one at a time by a single object that "makes our choices for us" (and removes the need in our second proof to use colorings with increasingly large numbers of colors). One possibility is to use ultrafilters. Although that is not the route that we will use to obtain our computability theoretic results, it is worth mentioning as a remark.

A *nonprincipal ultrafilter* on $\mathbb{N}$ is a collection $U$ of subsets of $\mathbb{N}$ such that the following conditions hold for all $X, Y \subseteq \mathbb{N}$. (The first three conditions make $U$ a filter; adding the fourth makes it an ultrafilter, while adding the fifth makes it nonprincipal.)

1. $\emptyset \notin U$.
2. If $X \in U$ and $X \subset Y$ then $Y \in U$.
3. If $X, Y \in U$ then $X \cap Y \in U$.
4. Either $X \in U$ or $\mathbb{N} \setminus X \in U$.
5. If $X$ is finite then $X \notin U$.

We think of the sets in $U$ as large. Note that all cofinite sets are in $U$ and, more generally, if $X \in U$ and $F$ is finite, then $X \setminus F \in U$. Note also that if $X_0, \ldots, X_k$ are pairwise disjoint sets partitioning some $Y \in U$, then exactly one $X_i$ is in $U$. To see that nonprincipal ultrafilters exist, let $F$ be collection of all cofinite subsets of $\mathbb{N}$. Then $F$ is a nonprincipal filter on $\mathbb{N}$ (i.e., it satisfies all of the above conditions except the fourth). The existence of an ultrafilter $U \supset F$ follows by Zorn's Lemma, and such a $U$ is necessarily nonprincipal.

We can use a nonprincipal ultrafilter $U$ to guide the choice of $I_{m+1}$ in our second proof of Ramsey's Theorem. Instead of defining $d$ as in that proof, we assume by induction that $I_m \in U$ and let

$$Z_{s,l} = \{x \in I_m \setminus \{a_m\} : c(s \cup \{x\}) = l\}.$$

Then for each $s \in F$, there is a unique $l_s$ such that $Z_{s,l_s} \in U$, and we can let $I_{m+1} = \bigcap_{s \in F} Z_{s,l_s} \in U$.

The ultrafilter used in this proof is not a countable object, but it is possible to replace it by a "sufficiently ultra" filter that can be built arithmetically (and hence, in particular, is countable), to get another proof of

$RT^n_{<\infty}$ in $ACA_0$. (See Towsner [203] for a discussion of this method and a way to deal with nonprincipal ultrafilters directly in reverse mathematics.) But our final proof of Ramsey's Theorem uses a more familiar object instead, and one about which we already have significant computability theoretic and reverse mathematical information, namely a finitely branching tree. It appeals to König's Lemma but can otherwise be carried out in $RCA_0$, and hence yields our easiest proof that $ACA_0 \vdash RT^n_{<\infty}$. In addition, its effectivization in Theorem 6.9 below will play a key role in our analysis of the computability theoretic strength of our versions of Ramsey's Theorem.

*Proof of Ramsey's Theorem, version 3.* Fix $n, k \geqslant 2$ and $c : [\mathbb{N}]^n \to k$. We show that there is an infinite, finitely branching tree $T$ such that each path on $T$ is an increasing function whose range is a prehomogeneous set for $c$. It then follows by König's Lemma that $c$ has an infinite prehomogeneous set.

Let $T$ be the tree of all sequences $m_0 < \cdots < m_{l-1}$ such that $\{m_0, \ldots, m_{l-1}\}$ is a prehomogeneous set for $c$ and for each $i < l$, we have that $m_i$ is the least $j > m_{i-1}$ with $c(s \cup \{j\}) = c(s \cup \{m_i\})$ for each $s \in [\{m_0, \ldots, m_{i-1}\}]^{n-1}$ (where we take $m_{-1} = -1$). The tree $T$ is finitely branching, since if $\sigma \in T$ then there is at most one immediate successor of $\sigma$ in $T$ for each function $[\text{rng}\,\sigma]^{n-1} \to k$. To see that it is also infinite, it is enough to show that for each $j$, it contains a $\sigma$ with $j \in \text{rng}\,\sigma$. So fix $j$ and consider the set $S$ of $\sigma \in T$ such that $j > \max \text{rng}\,\sigma$ and $\text{rng}\,\sigma j$ is a prehomogeneous set. The set $S$ is nonempty, since it contains the empty sequence, and is finite, so let $\sigma$ be an element of $S$ of maximal length. It is easy to check that $\sigma j \in T$. Finally, it is clear that the range of any path on $T$ is an infinite prehomogeneous set for $c$. $\qquad\square$

Other than the appeal to König's Lemma, the above proof can be carried out in $RCA_0$, so it yields another proof that for each $n$, we have $ACA_0 \vdash RT^n_{<\infty}$. It also allows us to improve on the previously noted fact that the double jump is enough to produce infinite prehomogeneous sets, in a way that will be quite useful in our computability theoretic analysis below.

**Theorem 6.9** (Jockusch [97]). *For each $n \geqslant 2$, every computable $k$-coloring of $[\mathbb{N}]^n$ has an infinite prehomogeneous set $A$ such that $A' \leqslant_T \emptyset''$.*

*Proof.* Let $c : [\mathbb{N}]^n \to k$ be computable. Then so is the tree $T$ in our third proof of Ramsey's Theorem. By the comments following Exercise 3.26, $T$

has a path $P$ such that $P' \leqslant_T \emptyset''$. Since this path is an increasing function, its range is $P$-computable. This range is the desired prehomogeneous set $A$.                                                                              □

It should be mentioned that is also possible to obtain this result from our second proof of Ramsey's Theorem, using the properties of PA degrees relative to $\emptyset'$ (and the fact that there is such a degree whose jump is $\emptyset''$, by the relativized form of the low basis theorem). For a discussion of these degrees that indicates how such a proof would go, see Section 6.5 below. In Theorem 6.21 we will see that Theorem 6.9 easily yields rough upper bounds on the complexity of infinite homogeneous sets.

Of course, instead of using the (relativized) low basis theorem in the proof of Theorem 6.9, we could use other basis theorems to obtain prehomogeneous sets with other desirable properties. In particular, using the cone-avoidance basis theorem we have the following fact, which will be useful below.

**Theorem 6.10** (Jockusch [97]). *For each $n \geqslant 2$ and each $X \nleqslant_T \emptyset'$, every computable $k$-coloring of $[\mathbb{N}]^n$ has an infinite prehomogeneous set $A$ such that $X \nleqslant_T A \oplus \emptyset'$.*

## 6.2   Ramsey's Theorem and the arithmetic hierarchy

Having established that $\mathrm{ACA}_0$ suffices to prove all our versions of Ramsey's Theorem other than full RT, the natural next question is whether we can do any better. And the natural first place to look for answers is in computability theoretic results. The first result in this direction was by Specker [199], who showed that there is a computable 2-coloring of $[\mathbb{N}]^2$ with no computable infinite homogeneous set, which implies that $\mathrm{RCA}_0 \nvdash \mathrm{RT}_2^2$. In his landmark paper [97], Jockusch improved this theorem as follows.

**Theorem 6.11** (Jockusch [97]). *There is a computable $c : [\mathbb{N}]^2 \to 2$ with no $\Sigma_2^0$ infinite homogeneous set.*

*Proof.* First note that it is enough to build $c$ to have no $\Delta_2^0$ infinite homogeneous set, since every infinite $\Sigma_2^0$ set has an infinite $\Delta_2^0$ subset (which is just the relativization to $\emptyset'$ of the fact that every infinite c.e. set has an infinite computable subset). Furthermore, there are uniformly computable total functions $h_0, h_1, \ldots : \mathbb{N} \times \mathbb{N} \to 2$ such that every $\Delta_2^0$ set is $\lambda n. \lim_s h_e(n, s)$

for some $e$. (Those unfamiliar with this fact should take it as an exercise. Note that some of the $h_e$ will not have limits.)

If $h_e$ does have a limit, call this limit $A_e$. We want to build $c$ so that for each $e$, if $A_e$ is defined and infinite, then there exist $w \neq x \in A_e$ and $y \neq z \in A_e$ such that $c(w, x) \neq c(y, z)$.

We proceed in stages. At stage $s$ we define $c(n, s)$ for all $n < s$. The stage is divided into substages $e = 0, 1, \ldots, s$. At each substage $e < s$ we define $c(n, s)$ for at most two $n$.

At substage $e < s$ proceed as follows. Let $B_{e,s} = \{n < s : h_e(n, s) = 1\}$. If $|B_{e,s}| < 2e + 2$ then do nothing. Otherwise, since we define $c(n, s)$ for at most two $n$ per substage, there exist $n_0 \neq n_1 \in B_{e,s}$ for which $c(n_i, s)$ has not been defined. Choose the least such $n_0, n_1$ and let $c(n_0, s) = 0$ and $c(n_1, s) = 1$.

At substage $s$, let $c(n, s) = 0$ for all $n < s$ such that $c(n, s)$ has not yet been defined.

Clearly $c$ is computable. Suppose that $A_e$ is defined and infinite. Let $m$ be such that $|A_e \upharpoonright m| = 2e + 2$. For all sufficiently large $s$, we have $B_{e,s} \upharpoonright m = A_e \upharpoonright m$. Pick such an $s > e$ in $A_e$. Then at substage $e$ of stage $s$, we define $c(n_0, s) = 0$ and $c(n_1, s) = 1$ for some $n_0, n_1 \in B_{e,s} \upharpoonright m \subset A_e$. Since $n_0, n_1, s \in A_e$, we thus ensure that $A_e$ is not a homogeneous set for $c$. $\square$

As there are $\omega$-models of $\mathrm{WKL}_0$ consisting entirely of low (and hence $\Delta_2^0$) sets, we have the following reverse mathematical consequence.

**Corollary 6.12** (Hirst [90]). $\mathrm{WKL}_0 \nvdash \mathrm{RT}_2^2$.

Thus, none of our versions of Ramsey's Theorem, except possibly $\mathrm{RT}_{<\infty}^1$, are provable in $\mathrm{WKL}_0$. Our next question then is whether we can obtain a reversal to $\mathrm{ACA}_0$, thus pinpointing the exact reverse mathematical strength of our principles. For triples and larger tuples, we can indeed do so.

**Theorem 6.13** (Jockusch [97]). *There is a computable $c : [\mathbb{N}]^3 \to 2$ such that any infinite homogeneous set for $c$ computes $\emptyset'$.*

*Proof.* For $x < s < t$, let $c(x, s, t) = 1$ if $\emptyset'[s] \upharpoonright x = \emptyset'[t] \upharpoonright x$, and $c(x, s, t) = 0$ otherwise. Suppose $H$ is an infinite homogeneous set for $c$. Let $x \in H$. Then there is a stage at which $\emptyset' \upharpoonright x$ stabilizes, so for all sufficiently large $s < t$, we must have $c(x, s, t) = 1$. Since $H$ is homogeneous, we have $c(x, s, t) = 1$ for all $x < s < t$ in $H$.

To compute $\emptyset'(e)$ from $H$, let $x < s \in H$ be such that $e < x$. Then $e \in \emptyset'$ iff $e \in \emptyset'[s]$, since otherwise there would be a $t > s$ in $H$ such that $\emptyset'[s] \restriction x \neq \emptyset'[t] \restriction x$, whence $c(x,s,t) = 0$. $\qquad\square$

Notice that the above argument works just as well if $H$ is prehomogeneous (indeed, every prehomogeneous set for the above coloring is homogeneous), which shows in particular that we cannot improve Theorem 6.9 to $A'' \leqslant_{\mathrm{T}} \emptyset''$.

 35 **Exercise 6.14** (Simpson [188]). Adapt the proof of Theorem 6.13 to show that $\mathrm{RCA}_0 + \mathrm{RT}_2^3 \vdash \mathrm{ACA}_0$.

It follows from Exercise 6.14 that all the principles we are considering are equivalent to $\mathrm{ACA}_0$ (and hence to each other) over $\mathrm{RCA}_0$, except possibly for $\mathrm{RT}^1_{<\infty}$, $\mathrm{RT}_2^2$, $\mathrm{RT}^2_{<\infty}$, and RT. We will consider $\mathrm{RT}^1_{<\infty}$ in Section 6.8, but note that this principle holds in every $\omega$-model of $\mathrm{RCA}_0$ (indeed, in any $\omega$-model at all, since it is a first order principle true of the standard natural numbers), so it cannot imply $\mathrm{ACA}_0$ (or any principle that does not hold in the $\omega$-model consisting of the computable sets) over $\mathrm{RCA}_0$. We will discuss RT in Section 6.3. Thus we are left with $\mathrm{RT}_2^2$ and $\mathrm{RT}^2_{<\infty}$. $\mathrm{RT}_2^2$ will be the main character of much of the rest of our tale. Many of our results for $\mathrm{RT}_2^2$ will apply to $\mathrm{RT}^2_{<\infty}$ as well, since these two principles have the same $\omega$-models, and we will discuss the situation for non-$\omega$-models in Section 6.8. But it is important to notice that the story of $\mathrm{RT}_2^n$ and $\mathrm{RT}^n_{<\infty}$ for $n \geqslant 3$ is not yet finished.

As discussed in Chapter 5, equivalence to $\mathrm{ACA}_0$ is a somewhat coarse classification. When we have shown that a principle of the form $\forall X \, [\theta(X) \rightarrow \exists Y \, \varphi(X,Y)]$ is equivalent to $\mathrm{ACA}_0$, it then becomes natural to find bounds on the complexity of $Y$ (in terms of the arithmetic hierarchy, at a first look) relative to that of $X$. The inductive nature of our proofs of Ramsey's Theorem suggests that the complexity of homogeneous sets should go up with $n$. As we will see, this is indeed the case.

In the terminology introduced in Section 3.2, Theorem 6.13 shows that the coding power of $\mathrm{RT}_2^3$ is at least at the $\emptyset'$ level. As $n$ increases, so does the amount of information we can encode into computable instances of $\mathrm{RT}_2^n$. We can show this using a theorem of Jockusch [97] that provides a tradeoff between the complexity of colorings and the size of tuples, and also allows us to extend Theorem 6.11 up the arithmetic hierarchy. (The last part of the following exercise will be useful in the proof of Theorem 6.17 below.)

36 **Exercise 6.15** (Jockusch [97]).

**a.** Show that if $d : [\mathbb{N}]^n \to 2$ is $\Delta_2^0$ then there is a computable $c : [\mathbb{N}]^{n+1} \to 2$ such that every infinite set that is homogeneous for $c$ is homogeneous for $d$.

**b.** Use this result and Theorem 6.13 to show that for each $n \geqslant 3$, there is a computable $c : [\mathbb{N}]^n \to 2$ such that any infinite homogeneous set for $c$ computes $\emptyset^{(n-2)}$.

**c.** Similarly, show that for each $n \geqslant 2$ there is a computable $c : [\mathbb{N}]^n \to 2$ with no $\Sigma_n^0$ infinite homogeneous set.

**d.** A 2-coloring $c$ is *unbalanced* if there is an $i < 2$ such that every infinite set that is homogeneous for $c$ is homogeneous to $i$. Show that, in part a, if $d$ is unbalanced then $c$ can be chosen to be unbalanced. Show also that, in part b, $c$ can be chosen to be unbalanced. Finally, show that if $c$ and $d$ are unbalanced 2-colorings of $[\mathbb{N}]^n$, then there is a 2-coloring $e$ of $[\mathbb{N}]^n$ such that any infinite set that is homogeneous for $e$ is homogeneous for both $c$ and $d$.

In Section 6.7 we will see that the result in part b of this exercise is in a sense tight, and has no nontrivial analog for $n = 2$. There is one way in which it can be improved, however. We begin with a strengthening of Theorem 6.13.

**Theorem 6.16** (Hirschfeldt and Jockusch [85]). *There is a computable 2-coloring of $[\mathbb{N}]^3$ such that any infinite prehomogeneous set has PA degree over $\emptyset'$.*

*Proof.* Here we assume each Turing functional $\Phi_e$ is $0, 1$-valued. Let $m < s < t$. If for all $e < m$ we have $\Phi_e^{\emptyset'[s]}(e)[s] = \Phi_e^{\emptyset'[t]}(e)[t]$ (which includes the possibility that both sides diverge) then let $c(m, s, t) = 1$. Otherwise, let $c(m, s, t) = 0$.

Suppose that $A$ is prehomogeneous for $c$ and fix $x$. Let $m \in A$ be such that $x < m$. Search for $s, t \in A$ such that $m < s < t$ and $\Phi_e^{\emptyset'[s]}(e)[s] = \Phi_e^{\emptyset'[t]}(e)[t]$ for all $e < m$. Such numbers must exist since there are only finitely many possibilities for the values of these computations. Let $g(x) = \Phi_x^{\emptyset'[s]}(x)[s]$ if the latter is defined, and otherwise let $g(x) = 0$. By prehomogeneity, $\Phi_x^{\emptyset'[s]}(x)[s] = \Phi_x^{\emptyset'[u]}(x)[u]$ for all $u > s$ in $A$, so if $\Phi_x^{\emptyset'}(x)\downarrow$ then $\Phi_x^{\emptyset'}(x) = \Phi_x^{\emptyset'[s]}(x)[s] = g(x)$. Thus $g$ is an $A$-computable completion of the partial function $e \mapsto \Phi_e^{\emptyset'}(e)$, and hence, as discussed in Section 3.1, $A \gg \emptyset'$. $\square$

In particular, there is a computable 2-coloring of $[\mathbb{N}]^3$ such that any infinite homogeneous set has PA degree over $\emptyset'$. We can extend this result to higher exponents.

**Theorem 6.17** (Hirschfeldt and Jockusch [85]). *Let $n \geqslant 3$. There is a computable 2-coloring of $[\mathbb{N}]^n$ such that any infinite homogeneous set has PA degree over $\emptyset^{(n-2)}$.*

*Proof.* We proceed by induction on $n \geqslant 3$ to prove a relativized form of the theorem: For each $X$, there is an $X$-computable 2-coloring of $[\mathbb{N}]^n$ such that for any infinite homogeneous set $H$, the set $X \oplus H$ has PA degree over $X^{(n-2)}$. The base case $n = 3$ is obtained by relativizing Theorem 6.16. Now fix $X$ and $n > 3$ and assume by induction that there is an $X'$-computable 2-coloring $d$ of $[\mathbb{N}]^{n-1}$ such that for any infinite homogeneous set $H$ for $d$, the set $X' \oplus H$ has PA degree over $X^{(n-2)}$.

By the relativized form of parts a and d of Exercise 6.15, there is an $X$-computable unbalanced 2-coloring $c$ of $[\mathbb{N}]^n$ such that if $H$ is homogeneous for $c$ then $H$ is homogeneous for $d$, and hence if $H$ is also infinite then $X' \oplus H$ has PA degree over $X^{(n-2)}$. By the relativized form of parts b and d of Exercise 6.15, there is also an unbalanced $X$-computable 2-coloring $e$ of $[\mathbb{N}]^n$ such that if $H$ is infinite and homogeneous for $e$ then $X \oplus H$ computes $X^{n-2}$, and hence computes $X'$.

By the relativized form of part d of Exercise 6.15, there is an $X$-computable 2-coloring $f$ of $[\mathbb{N}]^n$ such that if $H$ is infinite and homogeneous for $f$ then $H$ is homogeneous for both $c$ and $e$, which implies that $X' \oplus H$ has PA degree over $X^{(n-2)}$ and $X \oplus H$ computes $X' \oplus H$, and hence that $X \oplus H$ has PA degree over $X^{(n-2)}$.                                                  $\square$

The above theorem can also be proved using the following computability theoretic fact, where a function $g$ *majorizes* a function $f$ if $f(n) \leqslant g(n)$ for all $n$.

**Theorem 6.18** (Jockusch and McLaughlin [104]). *For each $k$, there is an increasing function $f \equiv_T \emptyset^{(k)}$ such that any function that majorizes $f$ computes $f$.*

$\boxed{37}$ **Exercise 6.19** (Hirschfeldt and Jockusch [85]). Prove Theorem 6.17 as follows.

(i) Fix $k$ and define a computable 2-coloring of $[\mathbb{N}]^3$ as follows. As in the proof of Theorem 6.16, assume each $\Phi_e$ is 0, 1-valued. By $\emptyset^{(k+1)}[s]$

we mean the stage $s$ of an $\emptyset^{(k)}$-computable approximation to $\emptyset^{(k+1)}$. Let $f$ be as in Theorem 6.18. Let $m < s < t$. If $s > f(m)$ and for all $e < m$ we have $\Phi_e^{\emptyset^{(k+1)}[s]}(e)[s] = \Phi_e^{\emptyset^{(k+1)}[t]}(e)[t]$ (which includes the possibility that both sides diverge), then let $c(m, s, t) = 1$. Otherwise, let $c(m, s, t) = 0$. Show that if $H$ is infinite and homogeneous for $c$, then $H \gg \emptyset^{(k+1)}$.

(ii) Proceed by induction on $n$, using part a as the base case, to show that for each $k$ and $n \geqslant 3$, there is an $\emptyset^{(k)}$-computable 2-coloring of $[\mathbb{N}]^n$ such that any infinite homogeneous set has PA degree over $\emptyset^{(k+n-2)}$.

We will see in Section 6.7 and the appendix beginning on Page 193 that Theorem 6.17 does not hold for $n = 2$. Note that being PA over $\emptyset^{(n-2)}$ is not in general sufficient to compute infinite homogeneous sets for colorings of $[\mathbb{N}]^n$, since there are $\Delta_n^0$ degrees that are PA over $\emptyset^{(n-2)}$, but by Exercise 6.15c, there are colorings of $[\mathbb{N}]^n$ with no $\Delta_n^0$ infinite homogeneous sets.

One consequence of Theorem 6.16 is that $\mathrm{KL} \leqslant_c \mathrm{RT}_2^3$, by the comments following Exercise 3.26, and in fact $\mathrm{KL} \leqslant_u \mathrm{RT}_2^3$ (where $\leqslant_c$ and $\leqslant_u$ are as defined in Section 2.2). On the other hand, Exercise 6.15c shows that $\mathrm{RT}_2^3 \not\leqslant_c \mathrm{KL}$, since every computable instance of $\mathrm{KL}$ has a $\Delta_3^0$ solution. We will see later that $\mathrm{RT}_2^2 \leqslant_c \mathrm{KL}$, but $\mathrm{KL} \not\leqslant_c \mathrm{RT}_2^2$.

Jockusch [97] showed that the result in Exercise 6.15c is tight, in the following sense.

**Theorem 6.20** (Jockusch [97]). *For each $n \geqslant 2$, every computable $k$-coloring of $[\mathbb{N}]^n$ has a $\Pi_n^0$ infinite homogeneous set.*

Another theorem giving upper bounds on the complexity of infinite homogeneous sets is the following, which will be improved in Theorem 6.55 below.

**Theorem 6.21** (Jockusch [97]). *For each $n \geqslant 1$, every computable $k$-coloring of $[\mathbb{N}]^n$ has an infinite homogeneous set $H$ such that $H' \leqslant_T \emptyset^{(n)}$.*

Both of these theorems can be proved inductively. As might be expected from the proofs in the previous section, the key to this induction is having a good computability theoretic upper bound on the complexity of prehomogeneous sets. Fortunately, we have already provided such a bound in Theorem 6.9.

*Proof of Theorem 6.21.* We prove the theorem in relativized form (which is necessary for the induction to carry through). The $n = 1$ case is trivial.

Now assume the $n$ case holds and let $c$ be a $k$-coloring of $[\mathbb{N}]^{n+1}$. By the relativized form of Theorem 6.9, $c$ has a prehomogeneous set $A$ such that $(c \oplus A)' \leqslant_T c''$. Let $d$ be the $k$-coloring of $[A]^n$ induced by $c$ (in the sense of Definition 6.5). By the inductive hypothesis, $d$ has an infinite homogeneous set $H$ such that $H' \leqslant_T d^{(n)} \leqslant_T (c \oplus A)^{(n)} \leqslant_T c^{(n+1)}$. This set is also homogeneous for $c$.                                                                          □

In the proof of Theorem 6.20, the induction has to start with $n = 2$ (as should be clear after solving the following exercise), and this base case is the difficult one. We will not give its proof here; it can be found in [97].

$\boxed{38}$ **Exercise 6.22.** Assume Theorem 6.20 holds for $n = 2$ and show that it holds in general.

We already have significant information on the strength of versions of Ramsey's Theorem. The above results reveal in particular that the inductive nature of the proofs in Section 6.1, and the resulting increase in complexity of the homogeneous sets produced for larger tuples, are not artifacts of a particular choice of proof technique, but something fundamental about the nature of Ramsey's Theorem. The ability to provide this kind of insight is one of the strengths of reverse mathematics and computable mathematics.

At this point we are left with at least three natural computability-theoretic questions (whose answers we have already foreshadowed above):

1. Can the upper bounds on the complexity of homogeneous sets given by Theorem 6.21 be improved?
2. Can the lower bounds on the coding power of homogeneous sets given by Exercise 6.15b be improved? Can we pinpoint the exact coding power of $\mathrm{RT}^n_2$ for $n \geqslant 3$?
3. Does $\mathrm{RT}^2_2$ have any coding power at all?

As we will see, the first two questions can be answered with inductive proofs, but these require more knowledge of the base case, namely $\mathrm{RT}^2_2$. In obtaining such knowledge, we will also be able to answer the last question, and determine where $\mathrm{RT}^2_2$ fits in the reverse mathematical universe.

In the next section, we discuss the strength of the full theorem RT. After that, we turn to the analysis of $\mathrm{RT}^2_2$ and related principles. We have already seen that $\mathrm{RT}^2_2$ is provable in $\mathrm{ACA}_0$ but not in $\mathrm{WKL}_0$. In Section 6.7, we will show that, unlike $\mathrm{RT}^n_2$ for $n \geqslant 3$, $\mathrm{RT}^2_2$ does not imply $\mathrm{ACA}_0$ over $\mathrm{RCA}_0$. We will also discuss the recent strengthening of this theorem by Liu [126], who showed that $\mathrm{RT}_2^2$ does not imply $\mathrm{WKL}_0$ over $\mathrm{RCA}_0$. But first we will see how $\mathrm{RT}_2^2$ can be thought of as the combination of two other principles, and how this fact gives us a useful tool in our analysis of $\mathrm{RT}_2^2$.

## 6.3 RT, ACA$_0'$, and the Paris-Harrington Theorem

We have mentioned that our proofs of $\mathrm{RT}_{<\infty}^n$ in $\mathrm{ACA}_0$ do not yield a proof of RT in $\mathrm{ACA}_0$, because of the external induction involved. Indeed, RT cannot be proved in $\mathrm{ACA}_0$. This fact follows from the following general theorem, which can be found in Wang [204], where it is said that it is "almost certainly a known theorem in proof theory." We give a model theoretic proof due to Jockusch.

Recall the comments on formalizing computability theory in $\mathrm{RCA}_0$, and in particular the definition of $Y = X'$, given in Section 4.2. Working in $\mathrm{ACA}_0$, for each set $X$ and each $k \in \omega$ (i.e., for each standard $k$), the $k$th jump $X^{(k)}$ exists (i.e., there exist $Y_0, \ldots, Y_k$ such that $Y_0 = X$ and $Y_{i+1} = Y_i'$). We write $Y \in \Sigma_k^{0,X}$ to mean that $Y$ is c.e. in $X^{(k)}$. Let $\mathcal{M} \vDash \mathrm{ACA}_0$ and let $M$ be the domain of the first order part of $\mathcal{M}$. Let $X$ be in the second order part of $\mathcal{M}$ and suppose that $Y$ is arithmetically definable over $X$, i.e., $Y = \{n \in M : \mathcal{M} \vDash \varphi(n)\}$ for some arithmetic formula $\varphi(x)$ with first order parameters from $M$ but no second order parameters other than $X$. Then the proof of Post's Theorem 2.3 shows that $\mathcal{M} \vDash Y \in \Sigma_k^{0,X}$ for some $k \in \omega$.

**Theorem 6.23** (see Wang [204]). *Let $\theta(X)$ and $\psi(X, Y)$ be arithmetic formulas with no free variables other than the ones displayed. If*

$$\mathrm{ACA}_0 \vdash \forall X \, [\theta(X) \rightarrow \exists Y \, \psi(X, Y)]$$

*then there is a $k \in \omega$ such that*

$$\mathrm{ACA}_0 \vdash \forall X \, [\theta(X) \rightarrow \exists Y \in \Sigma_k^{0,X} \, \psi(X, Y)].$$

*Proof.* Suppose that $\mathrm{ACA}_0 \vdash \forall X \, [\theta(X) \rightarrow \exists Y \, \psi(X, Y)]$. Add a second order constant symbol $C$ to the language of second order arithmetic. Let $T$ consist of $\mathrm{ACA}_0$ together with the sentence $\theta(C)$ and the sentences $\sigma_k \equiv \forall Y \in \Sigma_k^{0,C} \, \neg\psi(C, Y)$ for each $k \in \omega$. We claim that $T$ cannot be consistent.

Suppose otherwise and let $\mathcal{M} \vDash T$. Let $M$ be the domain of the first order part of $\mathcal{M}$, and let $S_M$ be the second order part of $\mathcal{M}$. Let $S_N$ consist of all subsets of $M$ that are arithmetically definable in our expanded language, i.e., of the form $\{n \in M : \mathcal{M} \vDash \varphi(n)\}$ for some arithmetic formula $\varphi(x)$ in the expanded language with no free variables other than $x$ and no second order parameters ($\varphi$ may have first order parameters from $M$). Notice that $S_N \subseteq S_M$, since $\mathcal{M} \vDash \text{ACA}_0$. Let $\mathcal{N}$ be the substructure of $\mathcal{M}$ with the same first order part as $\mathcal{M}$ and second order part $S_N$. We show that $\mathcal{N} \vDash \text{ACA}_0$.

Clearly $\mathcal{N}$ satisfies the basic axioms $P_0$ of Definition 4.1. Now let $\psi(x)$ be an arithmetic formula with parameters from $M \cup S_N$ and no free variables other than $x$. Let $Y_0, \ldots, Y_{l-1}$ be the second order parameters of $\psi$. Let $\varphi_0, \ldots, \varphi_{l-1}$ be arithmetic formulas in our expanded language defining $Y_0, \ldots, Y_{l-1}$, respectively. Replace all of the occurrences of atomic formulas $t \in Y_i$ in $\psi$ by $\varphi_i(t)$, to obtain a formula $\rho$. Then $\mathcal{N}$ satisfies both $\exists X \forall n [n \in X \leftrightarrow \rho(n)]$ (because $\{n \in M : \mathcal{M} \vDash \rho(n)\} \in S_N$) and $\forall n [\psi(n) \leftrightarrow \rho(n)]$, so it also satisfies $\exists X \forall n [n \in X \leftrightarrow \psi(n)]$.

Thus $\mathcal{N}$ is a model of $\text{ACA}_0$, and hence of $\forall X [\theta(X) \to \exists Y \psi(X,Y)]$. Since $\theta$ is arithmetic and $\mathcal{M} \vDash \theta(C)$, we also have $\mathcal{N} \vDash \theta(C)$, so there is a $Y \in S_N$ such that $\mathcal{N} \vDash \psi(C,Y)$. Since $\psi$ is arithmetic, $\mathcal{M} \vDash \psi(C,Y)$. But $Y$ has an arithmetic definition, so there is a $k \in \omega$ such that $\mathcal{M} \vDash Y \in \Sigma_k^{0,C}$, and hence $\mathcal{M} \vDash \neg \sigma_k$, contradicting the hypothesis that $\mathcal{M} \vDash T$.

Thus $T$ is inconsistent. Since $\sigma_{k+1}$ implies $\sigma_k$ for all $k$, there must be a $k$ such that $\text{ACA}_0 + \theta(C) + \sigma_k$ is inconsistent. That is, $\text{ACA}_0 \vdash \theta(C) \to \neg \sigma_k$, so by generalization on constants (see e.g. Enderton [53]), $\text{ACA}_0 \vdash \forall X [\theta(X) \to \exists Y \in \Sigma_k^{0,X} \psi(X,Y)]$.  $\square$

In particular, the above theorem implies that if $P$ is a principle expressed by a sentence of the form $\forall X [\theta(X) \to \exists Y \psi(X,Y)]$ with both $\theta$ and $\psi$ arithmetic, and $\text{ACA}_0 \vdash P$, then there is a $k \in \omega$ such that each computable instance of $P$ has a $\Sigma_k^0$ solution. By Exercise 6.15, there is no single $k$ such that for every $n$, every coloring of $[\mathbb{N}]^n$ has a $\Sigma_k^0$ infinite homogeneous set, so we have the following consequence.

**Corollary 6.24.** $\text{ACA}_0 \nvdash \text{RT}$.

Another interesting way to see that RT is stronger than $\text{ACA}_0$ over $\text{RCA}_0$ is via the Paris-Harrington Theorem [160], which gave the first "natural" example of a true arithmetic statement not provable in Peano Arithmetic. RT has the following finite analog, which we will denote by FRT: For each $n$, $k$, and $l$, there is a $u$ such that if $X$ is a finite set of size at least $u$, then every $k$-coloring of $[X]^n$ has a homogeneous set of size at least $l$. Paris and Harrington [160] gave the following strengthening of FRT, which we will denote by PH: For each $n$, $k$, and $l$, there is a $u$ such that if $X$ is a finite set of size at least $u$, then every $k$-coloring of $[X]^n$ has a homogeneous set $H$ of size at least $l$ such that $\min H \geqslant |H|$. Note that FRT and PH can be written as arithmetic sentences.

**39** **Exercise 6.25.** Show that $\mathrm{RCA}_0 + \mathrm{RT} \vdash \mathrm{PH}$.

FRT can be proved in $\Sigma_1^0$-PA, and hence in $\mathrm{RCA}_0$ (see Hájek and Pudlák [75]). Although PH seems quite similar to FRT, Paris and Harrington [160] showed that it cannot be proved in PA (see also [75] or Kaye [109]). As mentioned in Section 4.6, PA is the first order part of $\mathrm{ACA}_0$, so $\mathrm{ACA}_0$ does not prove PH. Combining this fact with Exercise 6.25 shows that RT is strictly stronger than $\mathrm{ACA}_0$ over $\mathrm{RCA}_0$. Indeed, we see that the first order part of $\mathrm{RCA}_0 + \mathrm{RT}$ is larger than that of $\mathrm{ACA}_0$. See [75] for more on the proof theoretic strength of finite versions of Ramsey's Theorem.

Exercise 6.15b shows that RT implies the existence of all finite iterates of the jump. This observation can be formalized by considering a system known as $\mathrm{ACA}_0'$, which is slightly more powerful than $\mathrm{ACA}_0$. This system is obtained by adding to $\mathrm{ACA}_0$ the statement that for all $X$ and all $n$, the $n$th jump of $X$ exists. Formally, $\mathrm{ACA}_0'$ consists of $\mathrm{ACA}_0$ together with the following statement: for each $X$ and each $n$, there is a sequence $X_0, \ldots, X_n$ such that $X_0 = X$ and $X_{i+1} = X_i'$ for all $i < n$.

**40** **Exercise 6.26.** Let $\varphi(x_0, \ldots, x_k, Y)$ be an arithmetic formula with no free variables other than the ones displayed. Show that $\mathrm{ACA}_0'$ proves that for each $X$ and each $n$, there is a sequence $X_0, \ldots, X_n$ such that $X_0 = X$ and $X_{i+1} = \{(m_0, \ldots, m_k) : \varphi(m_0, \ldots, m_k, X_i)\}$.

McAloon [133] noted that a proof of the following theorem can be extracted from that of a related theorem in Friedman, McAloon, and Simpson [64]. Other proofs can be found in Mileti [138], as a consequence of his computability theoretic analysis of another version of Ramsey's Theorem called the Canonical Ramsey Theorem; and in Dzhafarov and Hirst [48], in connection with yet another such version, called the Polarized Ramsey Theorem. We give a direct proof obtained by Dzhafarov [personal communication] as a simplification of the argument in [48].

**Theorem 6.27** (McAloon [133], Friedman, McAloon, and Simpson [64]). RT *is equivalent to* $\mathrm{ACA}_0'$ *over* $\mathrm{RCA}_0$.

*Proof.* We first argue in $\mathrm{ACA}_0'$. The third version of the proof of Ramsey's

Theorem on page 81 gives us a procedure that, given a coloring $c : [\mathbb{N}]^n \to k$ with $n, k \geqslant 2$, produces an infinite, finitely branching tree $T(c)$ such that the range of any path on $T(c)$ is an infinite prehomogeneous set for $c$. Let $d(c)$ be the coloring induced by the range of the leftmost path on $T(c)$ (in the sense of Definition 6.5). Then $d$ is an arithmetic operator. That is, there is an arithmetic formula $\varphi(x_0, x_1, Y)$ with no free variables other than the ones displayed such that $d(c) = \{(m, i) : \varphi(m, i, c)\}$ for all colorings $c$. (Writing out such a formula explicitly is a straightforward, if somewhat tedious, exercise.) Recall that, as noted following Definition 6.5, any set that is homogeneous for $d(c)$ is also homogeneous for $c$.

Now fix a coloring $c : [\mathbb{N}]^n \to k$. By Exercise 6.26, there is a sequence $c_0, \ldots, c_{n-1}$ such that $c_0 = c$ and $c_i = d(c_{i-1})$ for $0 < i < n$. By $\mathrm{RT}^1_{<\infty}$, the coloring $c_{n-1}$ has an infinite homogeneous set $H$. By (arithmetic) induction, $H$ is also homogeneous for each $c_i$, and in particular for $c$.

We now argue in $\mathrm{RCA}_0 + \mathrm{RT}$. Since $\mathrm{RT}$ implies $\mathrm{ACA}_0$, we are free to use arithmetic comprehension and arithmetic induction. Fix $X$ and $n$. For clarity of notation, we will denote $Y'[s]$ by $K_s^Y$. Define a family of finite sets as follows. Let $J^0(s_0) = X \restriction s_0$ and let

$$J^{k+1}(s_0, \ldots, s_{k+1}) = K_{s_1}^{J^k(s_1, \ldots, s_k)} \restriction s_0.$$

It is straightforward to check that arithmetic comprehension gives us a map taking $\langle k, s_0, \ldots, s_k \rangle$ to $J^k(s_0, \ldots, s_k)$. The idea is that the $J^k$-sets provide an approximation to the $k$th jump of $X$. We now define a coloring such that the numbers in any homogeneous set are large enough to pick out stages by which these approximations have settled.

Define $c : [\mathbb{N}]^{2n+1} \to n + 1$ by letting $c(s_0, s_1, \ldots, s_n, t_1, \ldots, t_n)$ be the least $k \in [1, n]$ such that $J^k(s_0, s_1 \ldots, s_k) \neq J^k(s_0, t_1 \ldots, t_k)$, if such a $k$ exists, and 0 otherwise. Let $H = \{h_0 < h_1 < \cdots\}$ be an infinite homogeneous set for $c$. Let $X_0 = X$ and let

$$X_k = \bigcup_i J^k(h_i, \ldots, h_{i+k}).$$

The sequence $X_0, \ldots, X_n$ exists by arithmetic comprehension, so it suffices to show that $X_{k+1} = X_k'$ for all $k < n$. We do so by (arithmetic) induction.

What we will actually show by induction is that $J^k(s_0, \ldots, s_k) = X_k \restriction s_0$ for all $s_0 < \cdots < s_k$ in $H$. At each step of the induction, we will also see that $X_{k+1} = X_k'$. Our hypothesis trivially holds for $k = 0$. Now assume it holds for $k < n$. Fix $s_0 < \cdots < s_k$ in $H$. Let $u$ be large enough so that $K_s^{X_k \restriction s} \restriction s_0 = K^{X_k} \restriction s_0$ for all $s \geqslant u$. If $t_1 < \cdots < t_k$ in $H$ are greater than $s_0$ and $u$, then

$$J^{k+1}(s_0, t_1, \ldots, t_{k+1}) = K_{t_1}^{J^k(t_1, \ldots, t_{k+1})} \restriction s_0 = K_{t_1}^{X_k \restriction t_1} \restriction s_0 = X_k' \restriction s_0.$$

Since the expression on the right does not depend on the $t_i$, the set $H$ cannot be homogeneous to $k + 1$, so

$$J^{k+1}(s_0, s_1, \ldots, s_{k+1}) = J^{k+1}(s_0, t_1, \ldots, t_{k+1}) = X'_k \upharpoonright s_0.$$

It now follows by definition that $X_{k+1} = \bigcup_i X'_k \upharpoonright h_i = X'_k$, as required. We also have $J^{k+1}(s_0, s_1, \ldots, s_{k+1}) = X_{k+1} \upharpoonright s_0$, so the inductive hypothesis holds for $k + 1$. $\square$

Also of interest is the system $\mathrm{ACA}_0^+$ obtained by adding to $\mathrm{ACA}_0$ the statement that for all $X$, the $\omega$th jump $X^{(\omega)} = \{\langle n, i \rangle : i \in X^{(n)}\}$ of $X$ exists. This system has come up for instance in the reverse mathematical analysis of Hindman's Theorem; see Section 10.3. $\mathrm{ACA}_0^+$ is clearly strictly stronger than $\mathrm{ACA}'_0$, since the arithmetic sets form an $\omega$-model of $\mathrm{ACA}'_0$ but not of $\mathrm{ACA}_0^+$.

## 6.4  Stability and cohesiveness

In their landmark paper on Ramsey's Theorem for Pairs, Cholak, Jockusch, and Slaman [20] introduced the important idea of splitting $\mathrm{RT}^2_2$ into a stable part and a cohesive part.

**Definition 6.28.** A coloring $c : [\mathbb{N}]^2 \to k$ is *stable* if $\lim_y c(x, y)$ exists for all $x$ (in other words, for each $x$, there is an $i < k$ such that $c(x, y) = i$ for almost all $y$).

*Stable Ramsey's Theorem for Pairs* ($\mathrm{SRT}^2_2$) is the statement that every stable 2-coloring of pairs of natural numbers has an infinite homogeneous set.

We can of course also define $\mathrm{SRT}^2_k$ in the same manner, but as with $\mathrm{RT}^2_2$, if $k > 2$ then $\mathrm{SRT}^2_k$ and $\mathrm{SRT}^2_2$ are equivalent. More interesting is $\mathrm{SRT}^2_{<\infty}$, the statement $\forall k \, \mathrm{SRT}^2_k$, which we will discuss in Section 6.8.

The following is a useful way to think of stable 2-colorings.

$\boxed{41}$ **Exercise 6.29** (Jockusch [97], Cholak, Jockush, and Slaman [20]). Show that for each computable stable $c : [\mathbb{N}]^2 \to 2$, there is a $\Delta^0_2$ set $A$ such that every subset of $A$ or $\overline{A}$ computes a homogeneous set for $c$. Conversely, show that for every $\Delta^0_2$ set $A$, there is a computable stable $c : [\mathbb{N}]^2 \to 2$ such that any homogeneous set for $c$ is contained in either $A$ or $\overline{A}$.

It is interesting to compare the above exercise to Exercise 6.15a. The natural way to solve the $n = 1$ case in Exercise 6.15a produces a stable coloring, and hence gives us the second half of Exercise 6.29, since a $\Delta_2^0$ 2-coloring of $\mathbb{N}$ is just the characteristic function of a $\Delta_2^0$ set. To obtain a converse (i.e., the first half of Exercise 6.29), we have to start with a stable coloring. We could define the concept of a stable 2-coloring $c$ of $[\mathbb{N}]^n$ as one such that for each $s \in [\mathbb{N}]^{n-1}$, there is an $i < 2$ for which $c(s \cup \{y\}) = i$ for almost all $y$. Then we would obtain a converse to Exercise 6.15a for stable colorings. The colorings in Exercise 6.15b can be taken to be stable, so these stable versions of Ramsey's Theorem for $n \geqslant 3$ are also equivalent to $\text{ACA}_0$, and hence to the corresponding general versions.

Recall that $X \subseteq^* Y$ means that $X \setminus Y$ is finite, and $\overline{X} = \mathbb{N} \setminus X$.

**Definition 6.30.** A set $C$ is *cohesive* for a collection of sets $R_0, R_1, \ldots$ if it is infinite and for each $i$, either $C \subseteq^* R_i$ or $C \subseteq^* \overline{R_i}$.

The *Cohesive Set Principle* (COH) is the statement that every countable collection of sets has a cohesive set.

We can use COH to transform any 2-coloring of pairs into a stable coloring, yielding the following result.

$\boxed{42}$ **Exercise 6.31** (Cholak, Jockusch, and Slaman [20]). Show that $\text{SRT}_2^2 + \text{COH}$ implies $\text{RT}_2^2$ over $\text{RCA}_0$.

In the converse direction, $\text{SRT}_2^2$ clearly follows from $\text{RT}_2^2$. So does COH, but the proof is somewhat subtle, because the most obvious proof uses more induction than is available in $\text{RCA}_0$. That proof goes as follows: Fix sets $R_0, R_1, \ldots$. By adding sets to this list, we may assume that for each $x \neq y$ there is an $i$ such that $R_i(x) \neq R_i(y)$. Let $i(x, y)$ be the least such $i$. Define a 2-coloring $c$ as follows. Given $x < y$, if $x \in R_{i(x,y)}$ then let $c(x, y) = 0$, and otherwise let $c(x, y) = 1$. Let $H$ be an infinite homogeneous set for this coloring. We claim that $H$ is cohesive for $R_0, R_1, \ldots$. Assume not, and let $k$ be least such that $H \cap R_k$ and $H \cap \overline{R_k}$ are both infinite. For $i < k$, let $F_i$ be whichever one of $H \cap R_i$ or $H \cap \overline{R_i}$ is finite, and let $F = \bigcup_{i<k} F_i$. Then $F$ is finite, and if $x \neq y$ are both not in $F$, then $i(x, y) \geqslant k$. Let $x, y, z \notin F$ with $x < y < z$ be such that $x, z \in H \cap R_k$ and $y \in H \cap \overline{R_k}$. Then $i(x, y) = i(z, y) = k$, so $c(x, y) = 0$ while $c(y, z) = 1$, contradicting the homogeneity of $H$.

This proof, which incidentally shows that $\text{COH} \leqslant_u \text{RT}_2^2$, goes through in $\text{RCA}_0$ except for two things. First, to say that there is a least $k$ such that $H \cap R_k$ and $H \cap \overline{R_k}$ are both infinite requires $\Sigma_2^0$-induction. Second, the fact that $F$ is finite because each $F_i$ with $i < k$ is finite is also not provable in $\text{RCA}_0$, as mentioned near the end of Section 4.1, though $\Sigma_2^0$-induction is also enough to prove it. So we do have a proof of COH in $\mathrm{RCA}_0 + \mathrm{I}\Sigma_2^0$, but it turns out that we can actually eliminate the uses of $\Sigma_2^0$-induction by being more careful.

**Theorem 6.32** (Mileti [137]; Jockusch and Lempp [unpublished]). $\mathrm{RT}_2^2$ *implies* COH *over* $\mathrm{RCA}_0$.

*Proof.* Given $R_0, R_1, \ldots$, define $c$ and $H$ as above. For simplicity of notation, we assume that $H$ is homogeneous to 0. If $H$ is homogeneous to 1 then the following proof works exactly the same, with the roles of 0 and 1 interchanged. Let $a_0 < a_1 < \cdots$ be the elements of $H$. We claim that for all finite sets $X$, if $R_k(a_n) = 1$ and $R_k(a_{n+1}) = 0$ for all $n \in X$, then $|X| < 2^k$. This claim suffices to prove the theorem, as it implies that for each $k$, there are fewer than $2^k$ many $n$ such that $R_k(a_n) = 1$ and $R_k(a_{n+1}) = 0$, and hence either $H \subseteq^* R_k$ or $H \subseteq^* \overline{R_k}$. The proof of the claim is by induction on $k$. Since we are quantifying over finite sets (which formally means quantifying over natural numbers that are codes for finite sets), this is an instance of $\Pi_1^0$-induction, which holds in $\mathrm{RCA}_0$.

If $R_0(a_n) = 1$ and $R_0(a_{n+1}) = 0$ then $c(a_n, a_{n+1}) = 1$, contradicting the fact that $H$ is homogeneous to 0. So there can be no such $n$, and thus our claim holds for $k = 0$. Now assume our claim holds for all numbers less than $k$. Suppose that $j < k$ and $Y$ is a finite set such that $R_j(a_n) = 0$ and $R_j(a_{n+1}) = 1$ for all $n \in Y$. Let $n_0 < \cdots < n_{l-1}$ be the elements of $Y$. For each $i < l - 1$, there is an $m_i \in (n_i, n_{i+1})$ such that $R_j(a_{m_i}) = 1$ and $R_j(a_{m_i+1}) = 0$. So by the inductive hypothesis, there are fewer than $2^j$ many such $i$. In other words, $|Y| \leqslant 2^j$. Now let $X$ be a finite set such that $R_k(a_n) = 1$ and $R_k(a_{n+1}) = 0$ for all $n \in X$. If $n \in X$ then we cannot have $i(a_n, a_{n+1}) = k$, as then $c(a_n, a_{n+1})$ would equal 1, so $i(a_n, a_{n+1}) < k$. For each $j < k$, let $Y_j = \{n \in X : i(a_n, a_{n+1}) = j\}$. If $n \in Y_j$ then $R_j(a_n) = 0$ and $R_j(a_{n+1}) = 1$, as otherwise $c(a_n, a_{n+1})$ would equal 1, so by the argument given above, $|Y_j| \leqslant 2^j$. But $X = \bigcup_{j<k} Y_j$, so $|X| \leqslant \sum_{j<k} 2^j < 2^k$. $\qquad\square$

We will analyze COH further in the next section, but for now we note the following facts, which are weaker forms of Theorem 6.45 below.

43 **Exercise 6.33** (Jockusch and Stephan [107]). Show that there is a sequence of uniformly computable sets $R_0, R_1, \ldots$ with no computable cohesive set. Thus $\mathrm{RCA}_0 \nvdash \mathrm{COH}$. [This exercise should be done before reading the following proof.]

**Theorem 6.34** (Jockusch and Stephan [107]; Cholak, Jockusch, and Slaman [20]). *There is a sequence of uniformly computable sets* $R_0, R_1, \ldots$ *with no low cohesive set. Thus* $\mathrm{WKL}_0 \nvdash \mathrm{COH}$.

*Proof.* If $\Phi_e^{\emptyset'[s]}(e)[s]\!\downarrow$ then let $R_e(s) = 1 - \Phi_e^{\emptyset'[s]}(e)[s]$. Otherwise, let $R_e(s) = 0$. Suppose that $C$ is cohesive for $R_0, R_1, \ldots$. Let $f(e) = 0$ if $C \subseteq^* \overline{R_e}$ and $f(e) = 1$ if $C \subseteq^* R_e$. We can compute $f(e)$ using $C'$ by searching for an $n$ such that either $m \in R_e$ for all $m > n$ in $C$ or $m \in \overline{R_e}$ for all $m > n$ in $C$. On the other hand, if $\Phi_e^{\emptyset'}(e)\!\downarrow = i$ then $R_e(s) = 1 - i$ for almost all $s$, so $f(e) = 1 - i \neq \Phi_e^{\emptyset'}(e)$. Thus $f \nleq_{\mathrm{T}} \emptyset'$. Since $f \leqslant_{\mathrm{T}} C'$, we have $C' \nleq_{\mathrm{T}} \emptyset'$. $\qquad\square$

It is also the case that $\mathrm{WKL}_0 \nvdash \mathrm{SRT}_2^2$. There are several ways to prove this fact. One is to use Theorem 6.37 below; another is to combine Theorems 6.81, 6.82, and 7.8 below. The following exercise gives yet another method. It requires knowledge of the finite injury priority method (see the proof of Theorem 9.11 below).

⎡44⎤ **Exercise 6.35** (Jockusch [97]). Show that there is a $\Delta_2^0$ set $A$ such that neither $A$ nor $\overline{A}$ has an infinite subset of hyperimmune-free degree. Conclude that $\mathrm{WKL}_0 \nvdash \mathrm{SRT}_2^2$. [Hint: Let $A_0 = \overline{A}$ and $A_1 = A$. For each $e$ and $i \in \{0, 1\}$ satisfy the requirement stating that if $\Phi_e$ is total then there is an $n$ such that the $n$th element of $A_i$ is greater than $\Phi_e(n)$. A single such requirement can be satisfied by choosing an $n$, waiting until $\Phi_e(n)\!\downarrow$, and then ensuring that all but $n - 1$ many numbers less than or equal to $\Phi_e(n)$ are in $A_{1-i}$. To combine the requirements, use a priority construction.]

We will see in Section 7.3 that COH does not imply $\mathrm{RT}_2^2$ over $\mathrm{RCA}_0$, and indeed that there is an $\omega$-model of $\mathrm{RCA}_0 + \mathrm{COH}$ that is not a model of $\mathrm{RT}_2^2$ (and hence not a model of $\mathrm{SRT}_2^2$). The relationship between $\mathrm{RT}_2^2$ and $\mathrm{SRT}_2^2$, on the other hand, has not yet been fully clarified. Computability theoretically, we know that $\mathrm{RT}_2^2 \nleq_{\mathrm{c}} \mathrm{SRT}_2^2$, since by Theorem 6.11, there is a computable instance of $\mathrm{RT}_2^2$ with no $\Delta_2^0$ solution, while by Exercise 6.29, every computable instance of $\mathrm{SRT}_2^2$ has a $\Delta_2^0$ solution. The question of whether $\mathrm{SRT}_2^2$ implies $\mathrm{RT}_2^2$ (or, equivalently, whether $\mathrm{SRT}_2^2$ implies COH) over $\mathrm{RCA}_0$, however, remained open and actively pursued for many years. It has recently been solved by Chong, Slaman, and Yang [24].

**Theorem 6.36** (Chong, Slaman, and Yang [24]). $\mathrm{SRT}_2^2$ *does not imply* $\mathrm{RT}_2^2$ *over* $\mathrm{RCA}_0$, *or even over* $\mathrm{WKL}_0$.

The proof of this theorem is a striking application of computability theory on nonstandard (first order) models. One proposal to separate $RT_2^2$ from $SRT_2^2$, made by Cholak, Jockusch, and Slaman [20], was to show that every $\Delta_2^0$ set has a low subset of either it or its complement (and hence every computable instance of $SRT_2^2$ has a low solution). Such a proof, if relativizable, would allow us to construct an $\omega$-model of $RCA_0 + SRT_2^2$ consisting entirely of low sets, which would then not be a model of $RT_2^2$. However, Downey, Hirschfeldt, Lempp, and Solomon [41] showed that this idea cannot work.

**Theorem 6.37** (Downey, Hirschfeldt, Lempp, and Solomon [41]). *There is a $\Delta_2^0$ set such that neither it nor its complement has an infinite low subset.*

To be more precise, this theorem shows that the above idea cannot work *for $\omega$-models*. Its proof uses an infinite injury priority construction that, as it turns out, cannot be carried out in $RCA_0 + B\Sigma_2^0$. Indeed, Chong, Slaman, and Yang [24] construct a first order structure $M \vDash P_0^- + B\Sigma_2^0$ with certain special features that make it possible to find, for each subset $A$ of the domain of $M$ that is $\Delta_2^0$ in the sense of $M$, an unbounded subset of $A$ or $\overline{A}$ that is low in the sense of $M$. Iterating this argument (and combining it with adding low solutions to WKL) then allows them to produce a model $\mathcal{M}$ of $WKL_0 + B\Sigma_2^0 + SRT_2^2$ whose second order part consists entirely of sets that are low in the sense of $M$. (We will see in Section 6.8 that in fact $RCA_0 \vdash SRT_2^2 \to B\Sigma_2^0$.) Theorem 6.36 then follows by combining this result with the following one, which shows that $\mathcal{M}$ is not a model of $RT_2^2$.

45 **Exercise 6.38** (Chong, Slaman, and Yang [24]). Show that the proof of Theorem 6.11 can be adapted to prove in $RCA_0 + B\Sigma_2^0$ that there is a computable 2-coloring of $[\mathbb{N}]^2$ with no infinite $\Delta_2^0$ homogeneous set.

Since the proof in [24] makes essential use of non-$\omega$-models, it does not answer the following question.

▶ **Open Question 6.39.** Is every $\omega$-model of $RCA_0 + SRT_2^2$ an $\omega$-model of $RT_2^2$? (We could also ask whether $SRT_2^2$ implies $RT_2^2$ over $RCA_0$ together with full induction, or at least some amount of extra induction, such as $I\Sigma_2^0$.)

We will discuss one possible approach to answering this question following Theorem 6.45 below.

## 6.5   Mathias forcing and cohesive sets

Although COH follows from Ramsey's Theorem, we give a direct proof
to introduce the notion of Mathias forcing. (See Section 2.3 for an introduction to forcing.) We will then see how this notion can be used to obtain significant information about the computability theoretic and reverse
mathematical strength of COH. In the following section, we will discuss
the implications for $RT_2^2$. As noted in Rogers [170], early computability
theoretic discussion of cohesiveness and related notions (including results
on the existence of cohesive sets) can be found in Dekker and Myhill [36]
and Rose and Ullian [171].

**Definition 6.40.** We define the notion of *Mathias forcing* $(P, \preccurlyeq)$. The set
$P$ consists of the pairs of the form $(F, I)$, where $F \subset \mathbb{N}$ is finite, $I \subseteq \mathbb{N}$
is infinite, and $\max F < \min I$. The idea is that the information about a
set $C$ embodied in $(F, I)$ is that $F \subseteq C$ and $C \setminus F \subseteq I$. Thus we define
$(F, I) \preccurlyeq (E, H)$ if $F \supseteq E$ and $I \subseteq H$, and $F \setminus E \subset H$. For a filter $\mathcal{F} \subset P$,
we think of $\bigcup_{(F,I) \in \mathcal{F}} F$ as our generic object, and define the forcing relation
$\Vdash$ as in Section 2.3.

**Proposition 6.41.** *Every sequence $R_0, R_1, \ldots \subseteq \mathbb{N}$ has a cohesive set.*

*Proof.* Let $D_n = \{(F, I) \in P : I \subseteq R_n \vee I \subseteq \overline{R_n}\}$ and let $E_n = \{(F, I) \in$
$P : |F| \geqslant n\}$. Each $D_n$ and $E_n$ is dense. (To see that $D_n$ is dense, let
$(F, H) \in P$. If $H \cap R_n$ is infinite, then let $I = H \cap R_n$. Otherwise, let $I =$
$H \cap \overline{R_n}$. Then $(F, I)$ extends $(F, H)$ and is in $D_n$.) Let $\mathcal{D}$ be the collection
of all $D_n$ and $E_n$, and let $G$ be a $\mathcal{D}$-generic filter. Let $C = \bigcup_{(F,I) \in G} F$. It
is easy to check that $C$ is infinite and cohesive for $R_0, R_1, \ldots$.            $\square$

If $R_0, R_1, \ldots$ are uniformly computable, then we can effectivize this
proof by replacing Mathias forcing with *computable Mathias forcing*, which
is defined in exactly the same way except that the infinite sets $I$ are required
to be computable. A computable Mathias condition $(F, I)$ may be indexed
by $\langle e, i \rangle$, where $e$ is the canonical index of $F$, and $i$ is an index for $I$ as a
computable set.

A crude effectivization of Proposition 6.41 can be obtained by noting
that $\emptyset''$ can decide whether a given computable set is infinite. This fact
means that, if $R_0, R_1, \ldots$ are uniformly computable and we replace Mathias
forcing by computable Mathias forcing, then the dense sets in the proof of
Proposition 6.41 are uniformly effectively dense relative to $\emptyset''$, as defined in

Section 2.3, and hence we can have $C \leqslant_T \emptyset''$, as discussed in that section. By using the technique of forcing the jump, as in Exercise 2.8, we can in fact get $C' \leqslant_T \emptyset''$, though it is also not difficult to do better than that:

| 46 | **Exercise 6.42** (Jockusch [96]).

a. Proceed as outlined above to show that if $R_0, R_1, \ldots$ are uniformly computable then they have a cohesive set $C$ such that $C' \leqslant_T \emptyset''$. [This exercise should be attempted before reading the proofs of the stronger versions below.]

b. Show that, in fact, if $X' \geqslant_T \emptyset''$, then there is an $X$-computable set that is cohesive for the collection of all c.e. sets. [In computability theory, such a set is simply called *cohesive*.]

Actually, we can do even better and get $C$ in the first part of the above exercise to be low$_2$, i.e., such that $C'' \leqslant_T \emptyset''$. There are two ways to do so. The most direct one is to control the double jump, i.e., to use $\emptyset''$-effective forcing to build a cohesive set $C$ while ensuring that $\emptyset''$ can compute $C''$. The difficulty with this idea is the following fact, which is noted in Cholak, Jockusch, and Slaman [20].

| 47 | **Exercise 6.43.** Working with the notion of computable Mathias forcing, show that there is a collection $\mathcal{D}$ that is uniformly effectively dense relative to $\emptyset''$ and such that any $\mathcal{D}$-generic set is high. [Hint: Use Theorem 2.2.]

Thus we need a modified notion of forcing to make our cohesive set low$_2$. A proof using such a notion was given by Cholak, Jockusch, and Slaman [20]. While this proof is slightly too long to include here, it is well worth studying. (One reason such double jump control proofs are interesting is that they can be converted into proofs of the kinds of conservativity results discussed in Section 6.8 and Chapter 7 below. Chong, Slaman, and Yang [23] remark that for a principle $P$ of the form $\forall X\,[\theta(X) \to \exists Y\,\psi(X,Y)]$, there is a "heuristic correspondence" between the computability theoretic conclusion that every instance $X$ of $P$ has a solution $Y$ such that $Y^{(n)} \leqslant_T X^{(n)}$ and the model theoretic one that we can extend a structure in which $\mathrm{I}\Sigma_n^0$ holds to add solutions to $P$ while preserving $\mathrm{I}\Sigma_n^0$, which yields conservativity results such as Theorem 6.83 below. See [20, 23, 177] for more on this subject; [23] in particular poses the interesting question of what the analog of the above heuristic for $\mathrm{B}\Sigma_n^0$ might be.)

The second technique for obtaining low$_2$ cohesive sets, which in fact establishes a stronger result, works by controlling the single jump instead of the double jump. The idea is to force the jump as in Exercise 6.42, but instead of working below $\emptyset'$, to work below a set $X$ that is low relative to $\emptyset'$ (i.e., $X' \leqslant_T \emptyset''$), thus ensuring that $C' \leqslant_T X$, whence $C'' \leqslant_T X' \leqslant_T \emptyset''$.

At first it might seem that there are no such sets $X$ powerful enough to guide our forcing argument. The issue, of course, is that to make the dense set $D_n$ effectively dense, it appears that we need to be able to decide, for a given infinite computable set $I$, whether $I \cap R_n$ is infinite. In general, such a question requires the full power of $\emptyset''$ to answer. However, answering that question is more than we actually need. It is enough to have an oracle that, given $I$ and $R_n$, tells us (correctly) either "$I \cap R_n$ is infinite" or "$I \cap \overline{R_n}$ is infinite". The point is that, if both of these sets are infinite, then the oracle may give either answer. So if the oracle's answer is "$I \cap \overline{R_n}$ is infinite", we still do not know whether $I \cap R_n$ is infinite, but we also do not *need* to know that to find an extension in $D_n$ of a condition involving $I$.

How much power is needed to obtain such an oracle? We can define a partial $\emptyset'$-computable function $\psi$ as follows. On input $i, j$, search for an $n$ and a $k = 0, 1$ such that $\forall m \geqslant n \, [\neg(\Phi_i(m)\!\downarrow = 1 \land \Phi_j(m)\!\downarrow = 1 - k)]$. If such numbers are found, then output $k$. Suppose that $g$ is a total $0, 1$-valued function extending $\psi$. Let $\Phi_i$ be an infinite set and $\Phi_j$ be a set. If $g(i, j) = 0$ then either $\psi(i, j) = 0$, in which case $\Phi_i \cap \Phi_j$ is finite, and hence $\Phi_i \cap \overline{\Phi_j}$ is infinite, or $\psi(i, j)\!\uparrow$, in which case again $\Phi_i \cap \overline{\Phi_j}$ is infinite. Similarly, if $g(i, j) = 1$ then $\Phi_i \cap \Phi_j$ is infinite. So any oracle computing such a $g$ is sufficient to carry out our forcing construction. By the relativized form of Theorem 3.21, if $X$ has PA degree relative to $\emptyset'$ (which recall is written as $X \gg \emptyset'$), then it can compute such a $g$. Note that in this case $\emptyset' \leqslant_T X$. By the relativized version of the low basis theorem, there is such an $X$ that is low relative to $\emptyset'$.

**Theorem 6.44** (Jockusch and Stephan [107]). *Let $X \gg \emptyset'$. If $R_0, R_1, \ldots$ are uniformly computable then they have a cohesive set $C$ such that $C' \leqslant_T X$. By choosing $X$ to be low relative to $\emptyset'$, we ensure that $C$ is low$_2$.*

*Proof.* Let $(P, \preccurlyeq)$ be the notion of computable Mathias forcing. If $\Phi_i$ is an infinite set, then let $c(\langle e, i \rangle) = (F, I)$ where $F$ is the finite set with canonical index $e$ and $I = \Phi_i$. We build our cohesive set $C$ as a sufficiently generic object. We need three families of conditions, the first two of which are the same as in Proposition 6.41. To ensure that $C$ is cohesive, let $D_n = \{(F, I) \in P : I \subseteq R_n \lor I \subseteq \overline{R_n}\}$, and to ensure that $C$ is infinite, let $E_n = \{(F, I) \in P : |F| \geqslant n\}$. Our third family of conditions forces the jump to ensure that $C' \leqslant_T X$. Let $J_e$ be the set of all $(F, I) \in P$ such that either

1. $\Phi_e^F(e)\downarrow$ with use at most $\min I$ or
2. $\Phi_e^{\widehat{F}}(e)\uparrow$ for all finite $\widehat{F}$ such that $F \subseteq \widehat{F} \subseteq F \cup I$.

Let $\mathcal{D}$ consist of all $D_n$, $E_n$, and $J_e$.

The $E_n$ are uniformly effectively dense. The $J_e$ are uniformly effectively dense relative to $\emptyset'$: Given $(F, I) \in P$, we can $\emptyset'$-computably determine (uniformly in $e$) whether there is a finite $E \subset I$ such that $\Phi_e^{F \cup E}(e)\downarrow$. If so, then let $\widehat{I}$ be the result of removing from $I$ all numbers $\leqslant \max E$ and all numbers less than the use of $\Phi_e^{F \cup E}(e)$. Then $(F \cup E, \widehat{I})$ is an extension of $(F, I)$ in $J_e$. Otherwise, $(F, I) \in J_e$.

The only place where we use the full power of $X$ is in showing that the $D_n$ are uniformly effectively dense relative to $X$. Let $g$ be an $X$-computable function such that if $\Phi_i$ is an infinite set and $\Phi_j$ is a set then either $g(i, j) = 1$ and $\Phi_i \cap \Phi_j$ is infinite or $g(i, j) = 0$ and $\Phi_i \cap \overline{\Phi_j}$ is infinite. Let $e_0, e_1, \ldots$ be a computable sequence of indices for $R_0, R_1, \ldots,$ respectively. Let $h$ be an $X$-computable function such that if $\Phi_i$ is an infinite set then $g(i, j) = 1 \rightarrow \Phi_{h(i,n)} = \Phi_i \cap \Phi_{e_n}$ and $g(i, j) = 0 \rightarrow \Phi_{h(i,n)} = \Phi_i \cap \overline{\Phi_{e_n}}$. For each $(F, \Phi_i) \in P$, the condition $(F, \Phi_{h(i,n)})$ is an extension of $(F, \Phi_i)$ in $D_n$.

Since $\emptyset' \leqslant_T X$, the collection $\mathcal{D}$ is uniformly effectively dense relative to $X$, so by Proposition 2.7, there is an $X$-computable sequence $i_0, i_1, \ldots$ such that $\{c(i_k) : k \in \mathbb{N}\}$ generates a $\mathcal{D}$-generic filter. Write $(F_k, I_k)$ for $c(i_k)$. Let $C = \bigcup_k F_k$. Then $C$ is cohesive for $R_0, R_1, \ldots.$

To see that $C' \leqslant_T X$, fix $e$. Note that $J_e$ is closed under extensions, so there is a $k$ such that $(F_k, I_k) \in J_e$. Since we can $\emptyset'$-computably recognize whether a given condition is in $J_e$, we can $X$-computably find such a $k$. Then $e \in C'$ iff $\Phi_e^{F_k}(e)\downarrow$, which again can be determined using $\emptyset'$, and hence using $X$. $\qquad\square$

Combining the relativized version of the above result with Proposition 4.27, we see that there is a model of $\mathrm{RCA}_0 + \mathrm{COH}$ consisting entirely of $\mathrm{low}_2$ sets. In particular, COH does not imply $\mathrm{ACA}_0$ over $\mathrm{RCA}_0$. In Corollary 6.51, we will see that COH does not imply even $\mathrm{WKL}_0$ over $\mathrm{RCA}_0$.

Let $Y$ be a set of low PA degree. We can relativize the above theorem to $Y$. Since $Y' \equiv_T \emptyset'$, we can still take $X$ to be low relative to $\emptyset'$. The advantage in this case is that, by Theorem 3.23, there is a uniformly

$Y$-computable sequence $R_0, R_1, \ldots$ containing all computable sets. A set $C$ is *r-cohesive* if it is cohesive for the collection of all computable sets. (Jockusch and Stephan [107] showed that the degrees of r-cohesive sets coincide with those of the cohesive sets defined in Exercise 6.42.) The above argument shows that if $X \gg \emptyset'$ then there is an r-cohesive set $C$ such that $C' \leqslant_T X$. The converse is also true. Indeed, there is even a collection of uniformly computable sets such that the jump of any cohesive set for this collection must have PA degree relative to $\emptyset'$. Thus we come to the interesting conclusion that COH behaves like WKL "one jump up" (though in a different way than KL; see the discussion before Exercise 3.27).

**Theorem 6.45** (Jockusch and Stephan [107]). *There are uniformly computable sets $A_0, A_1, \ldots$ such that the following are equivalent.*

1. $X \gg \emptyset'$.
2. *There is an r-cohesive set $C$ such that $C' \leqslant_T X$.*
3. *There is a cohesive set $C$ for $A_0, A_1, \ldots$ such that $C' \leqslant_T X$.*

*Proof.* We have already shown that 1 implies 2, and clearly 2 implies 3. Now, it is not difficult to show that there is a computable $0, 1$-valued function $f$ such that $\lim_s f(e, n, s) = \Phi_e^{\emptyset'}(n)$ if $\Phi_e^{\emptyset'}(n)\!\downarrow \leqslant 1$. Let $A_{\langle e,n \rangle} = \{s : f(e, n, s) = 1\}$. Let $C$ be a cohesive set for $A_0, A_1, \ldots$. Fix $e$ and $n$. Then $\lim_{s \in C} f(e, n, s)$ exists. Let $g(e, n) = \lim_{s \in C} f(e, n, s)$. Then $g \leqslant_T C'$, and for each $e$, the $C'$-computable function taking $n$ to $g(e, n)$ is a total $0, 1$-valued extension of $\Phi_e^{\emptyset'}$. Thus, by the relativized form of Theorem 3.21, $C' \gg \emptyset'$. $\qquad\square$

This theorem may have some bearing on Question 6.39. Recall that, by Theorem 6.37, there is a $\Delta_2^0$ set such that neither it nor its complement has an infinite low subset. The following question, however, is still open, and, as we show in Exercise 6.47 below, could lead to a solution to Question 6.39.

▶ **Open Question 6.46.** Does every $\Delta_2^0$ set have a subset of either it or its complement that is both low$_2$ and $\Delta_2^0$ (or at least both low$_n$ for some $n$ and $\Delta_2^0$)?

A set is 1-CEA($X$) if it is c.e. relative to $X$ and computes $X$. A set $A$ is $n$-CEA($X$) if there is a set $B$ that is $(n-1)$-CEA($X$) and such that $A$ is c.e. relative to $B$ and computes $B$. Jockusch, Lerman, Soare, and Solovay [102] extended Theorem 3.18 by showing that if $A$ is $n$-$\mathrm{CEA}(X)$ for some $n$ and has PA degree relative to $X$, then $A \geqslant_\mathrm{T} X'$.

[48] **Exercise 6.47** (Hirschfeldt, Jockusch, Kjos-Hanssen, Lempp, and Slaman [86]). Assume that every $\Delta_2^0$ set has a subset of either it or its complement that is both low$_n$ for some $n$ and $\Delta_2^0$, and that this fact is relativizable. Show that there is an $\omega$-model of $\mathrm{RCA}_0 + \mathrm{SRT}_2^2$ consisting entirely of sets whose jumps do not have PA degree relative to $\emptyset'$. Conclude that Question 6.39 has a negative answer.

The strongest negative answer to Question 6.46 would be that for every $X$, there is a set $A \leqslant_\mathrm{T} X'$ such that for every subset $S$ of $A$ or its complement, $(S \oplus X)'$ has PA degree relative to $X'$. If this were the case, we would have a strong positive answer to Question 6.39. Indeed, by Theorem 6.45, the above statement is equivalent to the statement that $\mathrm{COH} \leqslant_\mathrm{c} \mathrm{SRT}_2^2$.

▶ **Open Question 6.48.** Is $\mathrm{COH} \leqslant_\mathrm{c} \mathrm{SRT}_2^2$?

Dzhafarov [47] gave a partial result toward a negative answer to this question. More recently, Dzhafarov [personal communication] has also shown that $\mathrm{COH} \not\leqslant_\mathrm{u} \mathrm{SRT}_2^2$. Even if Question 6.48 does indeed have a negative answer, it is still possible that there is a set $A$ such that every $\Delta_2^0$ subset of $A$ or its complement is high. The strongest currently known positive result about $\Delta_2^0$ subsets of a $\Delta_2^0$ set or its complement is the following.

**Theorem 6.49** (Hirschfeldt, Jockusch, Kjos-Hanssen, Lempp, and Slaman [86]). *Let $A$ be $\Delta_2^0$ and let $C_0, C_1, \ldots$ be uniformly $\Delta_2^0$ noncomputable sets. Then there is an infinite $\Delta_2^0$ subset $X$ of either $A$ or $\overline{A}$ such that $C_i \not\leqslant_\mathrm{T} X$ for all $i$. In particular, every $\Delta_2^0$ set has an incomplete $\Delta_2^0$ subset of either it or its complement.*

We have seen in Theorem 6.34 that $\mathrm{WKL}_0 \nvdash \mathrm{COH}$. (Theorem 6.45 gives another proof of this fact, as the sets $A_0, A_1, \ldots$ cannot have a low cohesive set.) The converse is also true. The proof of this fact is an example of the important technique of proving that a principle $P$ does not imply another principle $Q$ over $\mathrm{RCA}_0$ by starting with an instance $I$ of $Q$ with no computable solution and adding solutions to instances of $P$ without adding a solution to $I$. We will see another in the proof of Seetapun's Theorem in Section 6.7.

**Theorem 6.50** (Cholak, Jockusch, and Slaman [20]). *Let $T$ be a computable infinite binary tree with no computable path, and let $C$ be the class* *of all sets that do not compute a path on $T$. Let the sets $A_0, A_1, \ldots$ be computable in an element $X$ of $\mathcal{C}$. Then $A_0, A_1, \ldots$ have a cohesive set $C$ such that $X \oplus C \in \mathcal{C}$. (We do not need to assume that $A_0, A_1, \ldots$ are uniformly $X$-computable, though that case would suffice to prove Corollary 6.51 below.)*

*Proof.* Let $X \in \mathcal{C}$ and let $A_0, A_1, \ldots$ be $X$-computable. Let $(P, \preccurlyeq)$ be the notion of $X$-computable Mathias forcing, defined as in Definition 6.40 but with the infinite sets $I$ required to be $X$-computable. We build our cohesive set $C$ as a sufficiently generic object. Although this proof uses $X$-computable Mathias forcing, it will not be an effective forcing construction.

As in the proofs of Proposition 6.41 and Theorem 6.44, we have the dense sets $D_n = \{(F, I) \in P : I \subseteq A_n \vee I \subseteq \overline{A_n}\}$ and $E_n = \{(F, I) \in P : |F| \geqslant n\}$ ensuring that $C$ is a cohesive set for $A_0, A_1, \ldots$. We need a third set of conditions ensuring that $X \oplus C$ does not compute a path on $T$.

Let $\Psi_0, \Psi_1, \ldots$ be a list of all Turing functionals with values in $2^{<\mathbb{N}}$ such that for any $Y$, if $\Psi_e^Y(n)\!\downarrow$ then $\Psi_e^Y(n)$ has length $n$, and if $\Psi_e^Y(m)\!\downarrow$ and $\Psi_e^Y(n)\!\downarrow$ for $m < n$ then $\Psi_e^Y(n)$ extends $\Psi_e^Y(m)$. Let $N_e$ be the set of all $(F, I) \in P$ for which there is an $n$ such that either

1. $\Psi_e^{X \oplus F}(n)\!\downarrow \,\notin T$ with use at most $\min I$ or
2. $\Psi_e^{X \oplus \widehat{F}}(n)\!\uparrow$ for all finite $\widehat{F}$ such that $F \subseteq \widehat{F} \subseteq F \cup I$.

If $C$ meets all of the $D_n$, $E_n$, and $N_e$ then it clearly has the required properties, so we are left with showing that each $N_e$ is dense.

Assume for a contradiction that $(F, I)$ has no extension in $N_e$. We claim that $T$ has an $X$-computable path, contradicting the choice of $X$. We define conditions $(F_0, I_0) \succcurlyeq (F_1, I_1) \succcurlyeq \cdots$ as follows. First, $X$-computably search for a finite $E \subset I$ such that $\Psi_e^{X \oplus (F \cup E)}(0)\!\downarrow$. (Note that then $\Psi_e^{X \oplus (F \cup E)}(0)$ must be the empty string, which is in $T$.) Such an $E$ must exist, as otherwise $(F, I)$ would itself be in $N_e$. Let $F_0 = F \cup E$, and let $I_0$ be the result of removing from $I$ all numbers $\leqslant \max E$ and all numbers less than the use of $\Psi_e^{X \oplus (F \cup E)}(0)$. Now suppose we are given $(F_{n-1}, I_{n-1})$ such that $\Psi_e^{X \oplus F_{n-1}}(n-1)\!\downarrow$ with use less than or equal to $\min I_{n-1}$. Then $X$-computably search for a finite $E \subset I_{n-1}$ such that $\Psi_e^{X \oplus (F_{n-1} \cup E)}(n)\!\downarrow$. Again, such an $E$ must exist. Let $F_n = F_{n-1} \cup E$, and let $I_n$ be the result of removing from $I_{n-1}$ all numbers $\leqslant \max E$ and all numbers less than the use of $\Psi_e^{X \oplus F_n}(n)$. Note that $\Psi_e^{X \oplus F_n}(n)$ must extend $\Psi_e^{X \oplus F_{n-1}}(n-1)$, by the definition of $\Psi_e$, and must be in $T$, as otherwise $(F_n, I_n)$ would be an extension of $(F, I)$ in $N_e$. It is easy to see that the $F_n$ are uniformly

$X$-computable, so that $\bigcup_n \Psi_e^{X \oplus F_n}(n)$ is an $X$-computable path on $T$. $\qquad\square$

By Proposition 4.27, there is an $\omega$-model of $\mathrm{RCA}_0 + \mathrm{COH}$ consisting entirely of sets in $\mathcal{C}$. This model is then not a model of $\mathrm{WKL}_0$, so we have the following result.

**Corollary 6.51** (Cholak, Jockusch, and Slaman [20]). $\mathrm{RCA}_0 + \mathrm{COH} \nvdash \mathrm{WKL}_0$.

## 6.6 Mathias forcing and stable colorings

The results of the previous section can help us obtain information about $\mathrm{RT}_2^2$ (and $\mathrm{RT}_{<\infty}^2$). We have seen that $\mathrm{COH} \leqslant_c \mathrm{RT}_2^2$, so lower bound theorems for COH apply to $\mathrm{RT}_2^2$ as well. In particular, Theorem 6.45 implies that there is a computable instance of $\mathrm{RT}_2^2$ all of whose solutions have jump of PA degree relative to $\emptyset'$. To obtain upper bound theorems for $\mathrm{RT}_2^2$ from ones for COH, we need analogous theorems for $\mathrm{SRT}_2^2$. Cholak, Jockusch, and Slaman [20] obtained such an analog to Theorem 6.44.

**Theorem 6.52** (Cholak, Jockusch, and Slaman [20]). *Every $\Delta_2^0$ set has an infinite $\mathrm{low}_2$ subset of either it or its complement.*

We will prove this theorem as a corollary to Theorem 6.57 below. Combining its relativized form with Theorem 6.44 and the solutions to Exercises 6.2c and 6.31 (keeping in mind also the fact that the class of $\mathrm{low}_2$ sets is closed under relativization), we have the following result.

**Theorem 6.53** (Cholak, Jockusch, and Slaman [20]). *Every computable $k$-coloring of $[\mathbb{N}]^2$ has a $\mathrm{low}_2$ infinite homogeneous set.*

Relativizing this theorem and applying Proposition 4.27, we obtain an $\omega$-model of $\mathrm{RT}_2^2$ (and hence of $\mathrm{RT}_{<\infty}^2$) consisting entirely of $\mathrm{low}_2$ sets, which establishes the following result.

**Corollary 6.54** (Seetapun, see [177]). $\mathrm{RT}_2^2$ *(or even $\mathrm{RT}_{<\infty}^2$) does not imply* $\mathrm{ACA}_0$ *over* $\mathrm{RCA}_0$.

We will discuss Seetapun's original route to this result in the next section.

Another consequence of Theorem 6.53 is an answer to the first question at the end of Section 6.2, whose proof is exactly the same as the proof of

Theorem 6.21 given above, except that our base case is now $n = 2$, and the inductive hypothesis gives us $H'' \leqslant_T d^{(n)}$.

**Theorem 6.55** (Cholak, Jockusch, and Slaman [20]). *For each $n \geqslant 2$, every computable $k$-coloring of $[\mathbb{N}]^n$ has an infinite homogeneous set $H$ such that $H'' \leqslant_T \emptyset^{(n)}$.*

Notice that this theorem is the best possible one along these lines, as any infinite homogeneous set $H$ for the 2-coloring of $[\mathbb{N}]^n$ in Exercise 6.15b has $H'' \geqslant_T \emptyset^{(n)}$.

Cholak, Jockusch, and Slaman [20] gave two proofs of Theorem 6.52. As in the case of Theorem 6.44, these are a double jump control proof and a single jump control proof. The latter establishes the stronger result that if $X \gg \emptyset'$, then every $\Delta_2^0$ set has an infinite subset of either it or its complement with an $X$-computable jump. Before presenting this proof, we give an auxiliary fact. Recall that a *lowness index* for a low set $B$ is an $e$ such that $\Phi_e^{\emptyset'} = B'$.

49 **Exercise 6.56.** Show that if $X \gg \emptyset'$ then there is an $X$-computable function $f$ such that if $e$ and $i$ are lowness indices for low sets $B$ and $C$, then $f(e, i) = 0$ implies that $B$ is infinite and $f(e, i) = 1$ implies that $C$ is infinite.

**Theorem 6.57** (Cholak, Jockusch, and Slaman [20]). *Let $X \gg \emptyset'$. Then every $\Delta_2^0$ set has an infinite subset of either it or its complement with an $X$-computable jump.*

*Proof.* Fix $X \gg \emptyset'$ and a $\Delta_2^0$ set $A$. We work with *low Mathias forcing*, which is defined in exactly the same way as Mathias forcing except that in each condition $(F, I)$, the infinite set $I$ is required to be low. The reason for working with low sets $I$ is the following. At one point we will need to obtain an extension of a condition $(F, I)$ by first forming an $I$-computable infinite binary tree, then letting $J$ be a path on this tree, then choosing either $I \cap J$ or $I \cap \overline{J}$ while making sure that the set we choose is infinite. Thus we need our $I$'s to belong to a class of sets that is closed under applications of some basis theorem for $\Pi_1^0$ classes, and is also simple enough that $X$ can make the above choice. The low sets fit both bills. Additionally, if $A$ has a low subset of either it or its complement, then we are done, so we may assume this is not the case. This assumption means that for every condition $(F, I)$, both $I \cap A$ and $I \cap \overline{A}$ are infinite. Let $(P, \preccurlyeq)$ be this notion of forcing. If $e$

is the canonical index of the finite set $F$ and $i$ is a lowness index for $I$ then let $c(\langle e, i \rangle) = (F, I)$.

Of course, our assumption means that we cannot expect to have our generic set be itself a subset of $A$ or $\overline{A}$. Instead, we will build a generic set $G$ such that both $G \cap A$ and $G \cap \overline{A}$ are infinite, and at least one of these sets has an $X$-computable jump.

Let $E_n = \{(F, I) \in P : |F \cap A| \geqslant n \land |F \cap \overline{A}| \geqslant n\}$. By our assumption on $A$, both $I \cap A$ and $I \cap \overline{A}$ are infinite, so each $E_n$ is dense. From a lowness index for $I$ and an $n$, we can $\emptyset'$-computably find elements $m_0 \in I \cap A$ and $m_1 \in I \cap \overline{A}$ such that $m_0, m_1 \geqslant n$. We can then also $\emptyset'$-computably determine a lowness index for $I \setminus [0, \max(m_0, m_1)]$. It follows that the $E_n$ are $\emptyset'$-effectively dense, and hence $X$-effectively dense.

Thus we are left with defining conditions to force the jump of either $G \cap A$ or $G \cap \overline{A}$. There is a standard computability theoretic trick we can use here. If we have two lists of requirements $R_0, R_1, \ldots$ and $Q_0, Q_1, \ldots$, and want to satisfy all the requirements in at least one of these lists, it is enough to satisfy the disjunctions $R_e \lor Q_i$ for all pairs $e, i$. Then if we do not satisfy all the $R_e$, let $e$ be such that $R_e$ is not satisfied. Since all $R_e \lor Q_i$ are satisfied, all $Q_i$ are satisfied.

Let $J_{e,i}$ be the set of all $(F, I) \in P$ satisfying at least one of the following conditions.

1. $\Phi_e^{F \cap A}(e) \downarrow$ with use at most $\min I$.
2. $\Phi_i^{F \cap \overline{A}}(i) \downarrow$ with use at most $\min I$.
3. There is no finite $D \subset I$ such that $\Phi_e^{(F \cap A) \cup D}(e) \downarrow$ and there is no finite $E \subset I$ such that $\Phi_i^{(F \cap \overline{A}) \cup E}(i) \downarrow$.

It might seem that in condition 3 we should quantify only over $D \subset I \cap A$ and $E \subset I \cap \overline{A}$, which would of course be enough to force the jumps of $G \cap A$ and $G \cap \overline{A}$ at $e$ and $i$, respectively. But then we would have little hope of having $J_{e,i}$ be $X$-effectively dense, since one-quantifier questions over $A$ generally require the full power of $A'$ to answer, and $A'$ could be as hard as $\emptyset''$. Of course, defining condition 3 as we do makes it a bit difficult to see why the $J_{e,i}$ should be dense at all, but we now argue that they are in fact $X$-effectively dense.

Let $(F, I) \in P$. Let $\mathcal{C}$ be the set of all $J \subseteq I$ for which there is no finite $D \subset J$ such that $\Phi_e^{(F \cap A) \cup D}(e) \downarrow$ and there is no finite $D \subset I \setminus J$ such that $\Phi_i^{(F \cap \overline{A}) \cup D}(i) \downarrow$. It is easy to check that $\mathcal{C}$ is a $\Pi_1^{0,I}$ class. First suppose that $\mathcal{C}$ is empty. Then in particular $I \cap A \notin \mathcal{C}$, so there is a $D$ such that either

1. $D \subset I \cap A$ and $\Phi_e^{(F \cap A) \cup D}(e)\downarrow$, in which case let $\widehat{I}$ be obtained from $I$ by removing all numbers less than the use of $\Phi_e^{(F \cap A) \cup D}(e)$; or

2. $D \subset I \cap \overline{A}$ and $\Phi_i^{(F \cap \overline{A}) \cup D}(i)\downarrow$, in which case let $\widehat{I}$ be obtained from $I$ by removing all numbers less than the use of $\Phi_i^{(F \cap \overline{A}) \cup D}(i)$.

In either case, $(F \cup D, \widehat{I})$ is in $J_{e,i}$, and we can use $\emptyset'$ to find such a $D$ and obtain an index (in the sense of the indexing function $c$) for $(F \cup D, \widehat{I})$ from one for $(F, I)$.

Now assume that $\mathcal{C} \neq \emptyset$. It is not difficult to see that a lowness index for a binary tree $T$ such that $\mathcal{C} = [T]$ can be obtained $\emptyset'$-computably from a lowness index for $I$. Thus, by Exercise 3.12, $\mathcal{C}$ has a low member $J$ such that a lowness index for $J$ can be obtained $\emptyset'$-computably from one for $I$. At least one of $J$ and $I \setminus J$ is infinite, and we can also $\emptyset'$-computably obtain a lowness index for $I \setminus J$. By Exercise 6.56, we can $X$-computably select an infinite set from among $J$ and $I \setminus J$. Let $\widehat{I}$ be the selected set. Then $(F, \widehat{I})$ is an extension of $(F, I)$ in $J_{e,i}$, and an index for $(F, \widehat{I})$ can be obtained $X$-computably from one for $(F, I)$.

Thus we see that the collection $\mathcal{D}$ of all $E_n$ and all $J_{e,i}$ is uniformly effectively dense relative to $X$, so by Proposition 2.7, there is an $X$-computable sequence $i_0, i_1, \ldots$ such that $\{c(i_k) : k \in \mathbb{N}\}$ generates a $\mathcal{D}$-generic filter. Write $(F_k, I_k)$ for $c(i_k)$. Let $G = \bigcup_k F_k$. Then $G \cap A$ and $G \cap \overline{A}$ are both infinite.

To see that either $(G \cap A)' \leqslant_{\mathrm{T}} X$ or $(G \cap \overline{A})' \leqslant_{\mathrm{T}} X$, fix $e$ and $i$. Since $J_{e,i}$ is closed under extensions, there is a $k_{e,i}$ such that $(F_{k_{e,i}}, I_{k_{e,i}}) \in J_{e,i}$. Since we can $\emptyset'$-computably recognize whether a given condition is in $J_{e,i}$, we can $X$-computably find such a $k_{e,i}$ and determine which of the three conditions in the definition of $J_{e,i}$ it satisfies. If for every $e$ there is an $i$ such that $(F_{k_{e,i}}, I_{k_{e,i}})$ satisfies either condition 1 or condition 3 in the definition of $J_{e,i}$, then by searching for such an $i$ for each $e$ we can $X$-computably determine $(G \cap A)'$. Otherwise, fix an $e$ with no such $i$. Then for every $i$, we have that $(F_{k_{e,i}}, I_{k_{e,i}})$ satisfies either condition 2 or condition 3 in the definition of $J_{e,i}$, and thus we can $X$-computably determine $(G \cap \overline{A})'$. $\square$

Let $X \gg \emptyset'$ and let $c$ be a computable 2-coloring of $[\mathbb{N}]^2$. By Theorem 4.26, there is a $Y$ such that $X \gg Y \gg \emptyset'$. By Theorem 6.44 and the solution to Exercise 6.31, there is a stable coloring $d$ such that $d' \leqslant_{\mathrm{T}} Y$ and any set homogeneous for $d$ is also homogeneous for $c$. By the relativized form of Theorem 6.57, $d$ has an infinite homogeneous set with $X$-computable jump. Thus we have the following result (again using Exercise 6.2c as well).

**Corollary 6.58** (Cholak, Jockusch, and Slaman [20]). *Let $X \gg \emptyset'$. Then every computable $k$-coloring of $[\mathbb{N}]^2$ has an infinite homogeneous set with $X$-computable jump.*

Recall that a computable instance of a principle $P$ is universal if any solution to it computes solutions to all computable instances of $P$. In Section 3.2, we saw examples of this phenomenon for WKL and KL. In light of the above corollary, the example we gave for KL shows that $RT_2^2 \leqslant_c KL$, but $KL \not\leqslant_c RT_2^2$. (However, $RT_2^2 \not\leqslant_u KL$, as we will see in Exercise 6.93 below.) As mentioned above, there is a computable 2-coloring of pairs all of whose infinite homogeneous sets have jumps of PA degree relative to $\emptyset'$. By the above corollary, we may call such a coloring a *jump universal* instance of $RT_2^2$. Mileti [137] showed that there is no universal instance of $RT_2^2$ or $SRT_2^2$ as a consequence of the following theorem, combined with Theorem 6.53.

**Theorem 6.59** (Mileti [137]). *Let $X$ be $low_2$. Then there is a $\Delta_2^0$ set with no infinite $X$-computable subset of it or its complement.*

The following exercise requires knowledge of the Recursion Theorem 2.1.

50 **Exercise 6.60** (Mileti [137]). Prove Theorem 6.59 as follows.

a. Let $A$ be $low_2$ and let $A_i = \{n : \langle i, n \rangle \in A\}$. Show that there is a $\Delta_2^0$ set $D$ such that for all $i$, if $A_i$ infinite then $A_i \not\subseteq D$ and $A_i \not\subseteq \overline{D}$. [Hint: We have requirements $R_{i,j}$ for $j \in \{0,1\}$ stating that if $A_i$ is infinite then there is an $n \in A_i$ such that $D(n) = 1 - j$. Use the $low_2$ness of $A$ and the recursion theorem relativized to $\emptyset'$ to show that there is a $0, 1$-valued function $g \leqslant_T \emptyset'$ such that $\lim_s g(i,j,s)$ exists for all $i, s \in \mathbb{N}$ and $j \in \{0,1\}$, and $R_{i,j}$ is satisfied iff $\lim_s g(i,j,s) = 1$. For each $s$, let $\langle i, j \rangle < s$ be least such that $g(i,j,s) = 0$ and let $D(s) = 1 - j$. (If no such $s$ exists then let $D(s) = 0$.) Verify that $D$ satisfies all the requirements.]

b. Use Theorem 3.23 to show that there is an $A$ that is low over $X$ such that every $X$-computable set is of the form $\{n : \langle i, n \rangle \in A\}$. Combine this result with the previous one to prove Theorem 6.59.

The contrast between Theorems 6.53 and 6.59 points to the important difference between saying that every computable instance of a principle $P$ has a solution computable in some set in a class $\mathcal{C}$, and saying that there is a single element of $\mathcal{C}$ that can compute solutions to all computable instances

*Slicing the Truth*

of $P$. If $P$ has a universal instance, the two statements are equivalent, but otherwise, the second can be much stronger.

Mileti [137] also showed that, in contrast to Theorem 6.49, if a $\Delta_2^0$ set $X$ can compute an infinite subset of any $\Delta_2^0$ set or its complement, then $X$ is complete.

## 6.7  Seetapun's Theorem and its extensions

As we have seen, for $n \geqslant 3$ we can code $\emptyset'$ into a computable coloring of $n$-tuples of natural numbers in a way that is recoverable from any infinite homogeneous set, from which fact we obtained the equivalence between $\mathrm{RT}_2^n$ and $\mathrm{ACA}_0$ for all $n \geqslant 3$. The question of whether $\mathrm{RT}_2^2$ implies $\mathrm{ACA}_0$ was settled by Seetapun (see Seetapun and Slaman [177]), via a computability theoretic result that shows that there is *no* noncomputable information that can be coded into a computable coloring of pairs in a way that is recoverable from any infinite homogeneous set, thus answering the third question at the end of Section 6.2. (Cf. the discussion of coding power in Section 3.2.)

**Theorem 6.61** (Seetapun, see [177]). *Let $c : [\mathbb{N}]^2 \to 2$ and let $X \not\leq_T c$. Then $c$ has an infinite homogeneous set $H$ such that $X \not\leq_T c \oplus H$.*

The original proof of Seetapun's Theorem in [177] shows that it also holds for countable collections of sets. That is, if $X_0, X_1, \ldots \not\leq_T c$ then there is an infinite homogeneous set $H$ such that $X_i \not\leq_T c \oplus H$ for all $i$.

Theorem 6.61 shows that we can apply Proposition 4.27 to the class of sets that do not compute $\emptyset'$ to give another proof of Corollary 6.54.

We give a newer proof of Theorem 6.61, considerably simpler than the original one, taken from Dzhafarov and Jockusch [50] (who also showed that the proof of Theorem 6.61 can be combined with that of Theorem 6.53 to construct infinite homogeneous sets for computable colorings of pairs that are both cone avoiding and low$_2$). This proof takes advantage of the decomposition of $\mathrm{RT}_2^2$ into $\mathrm{SRT}_2^2$ and COH. That is, we prove versions of Theorem 6.61 for each of these subprinciples, then combine them to prove Theorem 6.61 itself. We begin with the version for COH.

51  **Exercise 6.62** (Jockusch and Stephan [107]). Let $B$ be a set, let $A_0, A_1, \ldots$ be $B$-computable sets, and let $Y$ be such that $Y \not\leq_T B$. Show that $A_0, A_1, \ldots$ have a cohesive set $C$ such that $Y \not\leq_T B \oplus C$. [Hint: The proof is similar to that of Lemma 6.50 (using computable Mathias forcing

instead of $X$-computable Mathias forcing), but a bit simpler.]

We now prove a noncodability theorem for $\mathrm{SRT}_2^2$, as a corollary to the following result.

**Lemma 6.63** (Dzhafarov and Jockusch [50]). *Let $A \subseteq \mathbb{N}$ be any set and $X$ be noncomputable. Then there is an infinite subset of either $A$ or its complement that does not compute $X$.*

*Proof.* If $A$ is finite or cofinite then the theorem is clearly true, so we may assume that $A$ is infinite and coinfinite. It is enough to build a $G$ such that $G \cap A$ and $G \cap \overline{A}$ are both infinite, and for each $e$ and $i$,

$$X \neq \Phi_e^{G \cap A} \vee X \neq \Phi_i^{G \cap \overline{A}}. \tag{6.3}$$

If we have such a $G$ then either $G \cap A$ is an infinite subset of $A$ that does not compute $X$, or there is an $e$ such that $X = \Phi_e^{G \cap A}$, in which case $X \neq \Phi_i^{G \cap \overline{A}}$ for all $i$, so $G \cap \overline{A}$ is an infinite subset of $\overline{A}$ that does not compute $X$.

We use a variant of Mathias forcing in which in addition to the usual requirements on the conditions $(F, I)$, as given in Definition 6.40, we also require that $X \not\leq_{\mathrm{T}} I$. We may assume that for all such $I$, both $I \cap A$ and $I \cap \overline{A}$ are infinite, as otherwise a finite variant of $I$ is already a subset of either $A$ or its complement that does not compute $X$. (As we saw in the proof of Theorem 6.57, being able to make an assumption of this sort is often useful in forcing proofs like this one.)

If $G$ is sufficiently generic for this notion of forcing, then $G \cap A$ and $G \cap \overline{A}$ are both infinite, since for each $n$ the set of conditions $(F, I)$ such that $|F \cap A| > n$ and $|F \cap \overline{A}| > n$ is clearly dense. (Here we need to appeal to the assumption made in the previous paragraph.) To show that $G$ also satisfies (6.3) it is enough to prove the density of the set $N_{e,i}$ of conditions $(F, I)$ such that there is an $n$ for which every $G$ compatible with $(F, I)$ satisfies at least one of the following conditions:

1. $\Phi_e^{G \cap A}(n)\uparrow$,
2. $\Phi_i^{G \cap \overline{A}}(n)\uparrow$,
3. $\Phi_e^{G \cap A}(n)\downarrow \neq X(n)$, or
4. $\Phi_i^{G \cap \overline{A}}(n)\downarrow \neq X(n)$

(in other words, $(F, I)$ forces the disjunction of these conditions).

Fix a condition $(F, I)$. We show that $(F, I)$ has an extension in $N_{e,i}$.

*Case 1.* There are a finite $E \subseteq I \cap A$ and an $n$ such that $\Phi_e^{(F \cap A) \cup E}(n)\downarrow \neq X(n)$. Let $J$ be the result of removing from $I$ all numbers $\leq \max E$ and all numbers less than the use of $\Phi_e^{(F \cap A) \cup E}(n)$. Then $(F \cup E, J)$ is an extension of $(F, I)$ in $N_{e,i}$.

*Case 2.* Same as case 1 with $\overline{A}$ in place of $A$ and $i$ in place of $e$.

*Case 3.* Otherwise. For $n \in \mathbb{N}$ and $k = 0, 1$, let $P_{n,k}$ be the set of all $Z \subseteq I$ such that

1. there is no finite $E \subseteq Z$ such that $\Phi_e^{(F \cap A) \cup E}(n) \!\downarrow = k$ and
2. there is no finite $E \subseteq I \setminus Z$ such that $\Phi_i^{(F \cap \overline{A}) \cup E}(n) \!\downarrow = k$.

It is easy to check that the $P_{n,k}$ are uniformly $\Pi_1^{0,I}$ classes. Also, since cases 1 and 2 do not hold, for each $n$, we have $I \cap A \in P_{n, 1 - X(n)}$.

Let $S = \{(n, k) : P_{n,k} = \emptyset\}$. Then $S$ is $I$-c.e., by the relativized form of Proposition 3.9, and $(n, 1 - X(n)) \notin S$ for all $n$. There must be an $n$ such that $(n, X(n)) \notin S$, as otherwise we could compute $X$ from $I$. Fix such an $n$. By the relativized version of the Cone-Avoidance Basis Theorem 3.14, there is a $Z \in P_{n, X(n)}$ such that $I \oplus Z \not\geq_T X$. If $Z$ is infinite then let $J = Z$; otherwise let $J = I \setminus Z$. Note that in either case we have $J \not\geq_T X$. It is now easy to check that $(F, J) \in N_{e,i}$.                                    $\square$

**Corollary 6.64** (Seetapun, see [177]). *Let* $c : [\mathbb{N}]^2 \to 2$ *be stable and $B$-computable, and let* $X \not\leq_T B$. *Then $c$ has an infinite homogeneous set $H$ such that* $X \not\leq_T B \oplus H$.

*Proof.* Let $A$ be as in the relativized form of Exercise 6.29. By the relativized form of Lemma 6.63, there is an infinite subset $S$ of either $A$ or $\overline{A}$ such that $X \not\leq_T B \oplus S$. As in Exercise 6.29, we can $c$-computably, and hence $B$-computably, obtain a homogeneous set $H$ for $c$ from $S$. Then $B \oplus H \leq_T B \oplus S$, so $X \not\leq_T B \oplus H$.                                    $\square$

Note that in Lemma 6.63, we did not need to assume that $X \not\leq_T A$, but the assumption that $X \not\leq_T B$ in the above corollary is necessary, because the process of passing from $S$ to $H$ in its proof is $B$-computable, rather than computable.

We are now ready to prove Seetapun's Theorem.

*Proof of Theorem 6.61.* Let $c : [\mathbb{N}]^2 \to 2$ and let $X \not\leq_T c$. Let $A_m = \{n \neq m : c(m, n) = 1\}$. The sets $A_0, A_1, \ldots$ are $c$-computable, so by Exercise 6.62, $A_0, A_1, \ldots$ have a cohesive set $C$ such that $X \not\leq_T c \oplus C$. Let $d$ be the restriction of $c$ to $[C]^2$. Then $d$ is stable and $c \oplus C$-computable. Thus, by Corollary 6.64, $d$ has an infinite homogeneous set $H$ such that $X \not\leq_T c \oplus C \oplus H$. Then $H$ is also homogeneous for $c$, and $X \not\leq_T c \oplus H$.                                    $\square$

Using Seetapun's Theorem and an inductive construction like the one in the proof of Theorem 6.21, we obtain an answer to the second question at the end of Section 6.2, namely that the coding power of $RT^n_2$ for $n \geqslant 3$ is exactly at the $\emptyset^{(n-2)}$ level. There is a complication here that forces us to begin our induction with the $n = 3$ case, which we obtain by combining Seetapun's Theorem with Theorem 6.10. We will note why this is the case in the proof. (At the end of this section, we will discuss how this issue affects the proof of an extension of the following theorem, also given in [20].)

**Theorem 6.65** (Cholak, Jockusch, and Slaman [20]). *Let $n \geqslant 2$ and let $X \not\leqslant_T \emptyset^{(n-2)}$. Then every computable $k$-coloring of $[\mathbb{N}]^n$ has an infinite homogeneous set that does not compute $X$.*

*Proof.* We may assume $k = 2$. The $n = 2$ case is just Seetapun's Theorem. The $n = 3$ case is proved as follows. Let $c$ be a computable 2-coloring of $[\mathbb{N}]^3$ and let $X \not\leqslant_T \emptyset'$. By Theorem 6.10, $c$ has an infinite prehomogeneous set $A$ such that $X \not\leqslant_T A$. Let $d$ be the 2-coloring of $[A]^2$ induced by $c$ (in the sense of Definition 6.5). Then $d$ is $A$-computable, so by Seetapun's Theorem, $d$ has an infinite homogeneous set $H$ with $X \not\leqslant_T H$. This set is also homogeneous for $c$.

We now proceed by induction on $n$ to prove the relativized version of the theorem. Let $n \geqslant 3$ and assume the $n$ case holds. Let $c$ be a 2-coloring of $[\mathbb{N}]^{n+1}$ and let $X \not\leqslant_T c^{(n-1)}$. By the relativized form of Theorem 6.9, $c$ has a prehomogeneous set $A$ such $(c \oplus A)' \leqslant_T c''$. Let $d$ be the 2-coloring of $[A]^n$ induced by $c$. We have $(c \oplus A)^{(n-2)} \leqslant_T c^{(n-1)}$, so $X \not\leqslant_T (c \oplus A)^{(n-2)}$. (For $n = 2$ we do not necessarily have $(c \oplus A)^{(n-2)} \leqslant_T c^{(n-1)}$, which is why we had to start our induction with the $n = 3$ case.) Thus, by the inductive hypothesis, $d$ has an infinite homogeneous set $H$ such that $X \not\leqslant_T H \oplus c \oplus A$. This set is also homogeneous for $c$. $\qquad \square$

In particular, the coding power of $RT^3_2$, like that of KL, is exactly at the $\emptyset'$ level. There is an interesting difference between how these two principles can code $\emptyset'$, though. As we saw in Exercise 3.24, we can have a computable finitely branching tree with exactly one path such that this path has the same degree as $\emptyset'$. Thus there is a computable instance of KL for which every solution has exactly the degree of the halting problem (because there is only one solution). Clearly, this can never be the case for $RT^3_2$, since any subset of a homogeneous set is homogeneous, and any infinite set has continuum many infinite subsets, while any degree is countable. We might then ask, for instance about the coloring in the proof of Theorem 6.13, which degrees contain infinite homogeneous sets for that coloring. Clearly the degree of $\emptyset'$ does, since if we let $s_0 = 0$ and $s_{n+1}$ be least such that $\emptyset'[s_{n+1}] \upharpoonright s_n = \emptyset' \upharpoonright s_n$, then $\{s_0, s_1, \ldots\}$ is a homogeneous set for that coloring of the same degree as $\emptyset'$; and no degree that is not above that of $\emptyset'$ does, by the theorem itself. We can now get a full answer to our question from the following remarkably general theorem, which applies to homogeneous sets, cohesive sets, and many other classes of objects.

**Theorem 6.66** (Jockusch [99]). *Let $C$ be a collection of infinite sets that contains an arithmetic set, such that if $A \subset B$ is infinite and $B \in C$ then $A \in C$. Then the class of degrees of elements of $C$ is closed upwards.*

It is also natural to ask whether all sets can be coded in some (not necessarily computable) instance of RT. Perhaps surprisingly, the answer is negative, and indeed we can use computability theoretic results to give an exact characterization of which sets can be so coded. A set $X$ is *encodable* if every infinite set has a subset that computes $X$.

**Theorem 6.67** (Solovay [198]). *A set is encodable iff it is hyperarithmetic.*

Theorem 6.18 generalizes to transfinite jumps, and indeed we have the following characterization.

**Theorem 6.68** (Jockusch and McLaughlin [104]). *A set is hyperarithmetic iff it is computed by an increasing function $f$ such that any function that majorizes $f$ computes $f$.*

**Corollary 6.69** (Hirschfeldt and Jockusch [85]). *Let $n, k \geq 2$. Then a set $X$ is hyperarithmetic iff there is a $k$-coloring of $[\mathbb{N}]^n$ such that every infinite homogeneous set computes $X$.*

*Proof.* First suppose there is a $k$-coloring $c$ of $[\mathbb{N}]^n$ such that every infinite homogeneous set computes $X$. Given an infinite set $A$, let $d$ be the restriction of $c$ to $[A]^n$. Let $H$ be homogeneous for $d$. Then $H \subseteq A$, and $H$ is also homogeneous for $c$, and hence computes $X$. Thus $X$ is encodable, and hence hyperarithmetic.

Now suppose $X$ is hyperarithmetic, and let $f \geq_T X$ be as in Theorem 6.68. For $x_0 < \cdots < x_{n-1}$, let $c(x_0, \ldots, x_{n-1}) = 1$ if $x_1 \geq f(x_0)$, and let $c(x_0, \ldots, x_{n-1}) = 0$ otherwise. Let $H$ be homogeneous for $c$. Then it is easy to see that $H$ must be homogeneous to 1. We may assume that the least element of $H$ is greater than or equal to $f(0)$. Let $g(n)$ be the

$(n+1)$st element of $H$ in natural number order. Then $g$ majorizes $f$, so we have $X \leqslant_\mathrm{T} f \leqslant_\mathrm{T} g \leqslant_\mathrm{T} H$. □

Note that the above result does not hold for $n = 1$:

| 52 | **Exercise 6.70** (Dzhafarov and Jockusch [50]). Show that if $c$ is a $k$-coloring of $\mathbb{N}$ and $X$ is not computable, then there is a homogeneous set for $c$ that does not compute $X$. [Hint: See Lemma 6.62.]

Since the strength of $\mathrm{RT}_2^2$ is strictly in between $\mathrm{RCA}_0$ and $\mathrm{ACA}_0$, the natural next step is to compare it with $\mathrm{WKL}_0$. We have seen in Corollary 6.12 that $\mathrm{WKL}_0$ does not imply $\mathrm{RT}_2^2$. Whether $\mathrm{RT}_2^2$ implies $\mathrm{WKL}_0$ was for a long time one of the major open questions in reverse mathematics. It was recently solved by Liu [126].

**Theorem 6.71** (Liu [126]). $\mathrm{RT}_2^2$ *does not imply* $\mathrm{WKL}_0$ *over* $\mathrm{RCA}_0$.

As with Seetapun's Theorem, Liu's Theorem is proved using an $\omega$-model, so it applies to $\mathrm{RT}_{<\infty}^2$ as well. Its proof shows that if $c$ is a 2-coloring of $[\mathbb{N}]^2$ that is not of PA degree, then $c$ has an infinite homogeneous set $H$ such that $c \oplus H$ is not of PA degree. Thus we see that Theorem 6.17 does not hold for $n = 2$.

Liu's proof uses a forcing argument like the ones we have discussed, but with an intricate elaboration on Mathias forcing. I have avoided including long proofs in this book, in the hope that readers will consult the original papers. However, it is probably worth including at least one example of a longer and more combinatorially complicated argument in reverse mathematics. Thus a proof of Liu's Theorem can be found in the appendix that begins on Page 193.

In the follow-up paper [127], Liu extended his method to obtain several interesting corollaries in reverse mathematics and algorithmic randomness. In particular, he proved the following strengthening of Theorem 6.71.

**Theorem 6.72** (Liu [127]). $\mathrm{RT}_2^2$ *does not imply* $\mathrm{WWKL}_0$ *over* $\mathrm{RCA}_0$.

Theorems 6.12 and 6.71 show that $\mathrm{RT}_2^2$ and $\mathrm{WKL}_0$ are *independent* over $\mathrm{RCA}_0$. One might thus wonder about their combined power. As pointed out in Exercise 4.28, one of the advantages of using Proposition 4.27 to build $\omega$-models is the ease with which we can combine constructions, as long as they use the same collection $\mathcal{C}$ in the statement of the proposition. In this case, we can either combine Theorem 6.53 with the low basis theorem and use Exercise 4.28 to obtain an $\omega$-model of $\mathrm{WKL}_0 + \mathrm{RT}_2^2$ consisting entirely of low$_2$ sets, or combine Theorem 6.61 with the cone-avoidance basis theorem and use Exercise 4.28 to obtain an $\omega$-model of $\mathrm{WKL}_0 + \mathrm{RT}_2^2$ that does not contain $\emptyset'$. In either case, we have the following stronger form of Corollary 6.54.

**Theorem 6.73** (Seetapun, see [177]). $\mathrm{RT}_2^2$ *does not imply* $\mathrm{ACA}_0$ *over* $\mathrm{WKL}_0$.

As we will see in the appendix beginning on Page 193, the proof of Liu's Theorem in [126] shows that for any set $A$, there is an infinite subset of $A$ or $\overline{A}$ that does not have PA degree, which relativizes to show that if the infinite set $X$ does not have PA degree, then for any set $A$, there is an infinite subset of $X \cap A$ or $X \cap \overline{A}$ that does not have PA degree. Wang [206] used this fact to prove the following result, which should be compared with Theorem 6.45.

[53] **Exercise 6.74** (Wang [206]). Show that every sequence $A_0, A_1, \ldots$ has a cohesive set that is not of PA degree. Note that there is no effectivity assumption on the $A_i$. [Hint: Use Mathias forcing with conditions $(F, I)$ where $I \not\gg \emptyset$.]

We finish this section by discussing an extension of Theorem 6.65. The following fact is stated as Theorem 12.2 in Cholak, Jockusch, and Slaman [20].

**Theorem 6.75** (Cholak, Jockusch, and Slaman [20]). *Let* $n \geqslant 2$ *and* $C_0, C_1, \ldots$ *be sets such that* $C_i \not\leqslant_\mathrm{T} \emptyset^{(n-2)}$ *for all* $i$. *Then every computable* $k$-*coloring of* $[\mathbb{N}]^n$ *has an infinite homogeneous set* $H$ *such that* $H' \not\geqslant_\mathrm{T} \emptyset^{(n)}$ *and* $C_i \not\leqslant_\mathrm{T} H$ *for all* $i$.

However, there is a flaw in the proof in [20]. As in the case of Theorem 6.65, the proof is by induction. The base case $n = 2$ is correct, but the $n = 3$ case does not follow from it, for the same reason as in the proof of Theorem 6.65. Thus the $n = 3$ case must be proved separately, after which the induction proceeds as in [20]. We give a proof of the $n = 3$ case, based on the proof of the $n = 2$ case in [20] and due to Hirschfeldt and Jockusch [85]. One reason for including this proof here is an example of how deep facts about the structure of the Turing degrees can be used to prove theorems in computable mathematics. We thus begin by stating three such facts that will be used in the proof.

We write $\deg(X)$ for the Turing degree of $X$ and use standard computability theoretic notation for Turing degrees: Degrees are denoted by boldface lowercase letters. The degree of computable sets is $\mathbf{0}$. Let $A \in \mathbf{a}$ and $B \in \mathbf{b}$. Then $\mathbf{a}' = \deg(A')$ and $\mathbf{a} \vee \mathbf{b} = \deg(A \oplus B)$, and $\mathbf{a} \leqslant \mathbf{b}$ iff $A \leqslant_T B$. (All of these definitions are clearly independent of the choice of $A$ and $B$.) We say that $\mathbf{d_0}, \mathbf{d_1}, \dots$ are *uniformly $\mathbf{a}$-computable* if there are $D_0 \in \mathbf{d_0}, D_1 \in \mathbf{d_1}, \dots$ such that $\{\langle i, n \rangle : n \in D_i\}$ is $\mathbf{a}$-computable (i.e., computable relative to $A \in \mathbf{a}$, which again does not depend on the choice of $A$).

Let $I$ be an ideal in the Turing degrees. Degrees $\mathbf{f}$ and $\mathbf{g}$ are an *exact pair* for $I$ if $I = \{\mathbf{a} : \mathbf{a} \leqslant \mathbf{f} \text{ and } \mathbf{a} \leqslant \mathbf{g}\}$.

**Theorem 6.76** (Kleene and Post [111]; Lacombe [119]; Spector [200]). *Let $I = \{\mathbf{d_0}, \mathbf{d_1}, \dots\}$ be an ideal in the Turing degrees. Then $I$ has an exact pair $\mathbf{f}, \mathbf{g}$. Furthermore, these degrees can be chosen so that the $\mathbf{d_i}$ are both uniformly $\mathbf{f}$-computable and uniformly $\mathbf{g}$-computable.*

A proof of Theorem 6.76 can be found in the proof of Theorem V.4.3 of Odifreddi [158]. The last statement in the above theorem is not explicitly stated there, but follows from the referenced proof.

**Theorem 6.77** (Friedberg [58]). *If $\mathbf{d} \geqslant \mathbf{0}'$ then there is an $\mathbf{a}$ such that $\mathbf{d} = \mathbf{a}'$.*

Theorem 6.77 was the first of several jump inversion theorems. The following is another; others can be found in [40, 158, 196], for instance.

**Theorem 6.78** (Posner and Robinson [163]). *Let $\mathbf{g} \geqslant \mathbf{0}'$ and let $\mathbf{e_0}, \mathbf{e_1}, \dots$ be nonzero and uniformly $\mathbf{g}$-computable. Then there is a $\mathbf{d}$ such that $\mathbf{g} = \mathbf{d}' = \mathbf{d} \vee \mathbf{e_i}$ for all $i$.*

We will also need the following exercise.

54 **Exercise 6.79** (Hirschfeldt and Jockusch [85]). Use Corollary 6.58 and induction to show that if $X \gg \emptyset^{(n-1)}$ then every computable $k$-coloring of $[\mathbb{N}]^n$ has an infinite homogeneous set with $X$-computable jump.

*Proof of the $n = 3$ case of Theorem 6.75.* We can assume $k = 2$. Let $c$ be a 2-coloring of $[\mathbb{N}]^3$. Let $\mathbf{c_i} = \deg(C_i)$. Let $\mathbf{d_0} = \mathbf{0}''$. If $\mathbf{d_i} \vee \mathbf{c_i} \not\geqslant \mathbf{0}'''$ then let $\mathbf{e_i} = \mathbf{c_i}$; otherwise let $\mathbf{e_i} = \mathbf{0}''$. Let $\mathbf{d_{i+1}} = \mathbf{d_i} \vee \mathbf{e_i}$. By Theorem 6.76, the ideal generated by the $\mathbf{d_i}$ has an exact pair $\mathbf{f}, \mathbf{g}$ such that the $\mathbf{e_i}$ are uniformly computable in both $\mathbf{f}$ and $\mathbf{g}$. Since $\mathbf{0}'''$ is not in this ideal, at least one of $\mathbf{f}$ and $\mathbf{g}$ is not above $\mathbf{0}'''$, say $\mathbf{g} \not\geqslant \mathbf{0}'''$. Note that, since $\mathbf{d_0} = \mathbf{0}''$, we have $\mathbf{g}' \geqslant \mathbf{0}'''$. By Theorem 6.78 relativized to $\mathbf{0}'$, there is a $\mathbf{d} \geqslant \mathbf{0}'$ such that $\mathbf{g} = \mathbf{d}' = \mathbf{d} \vee \mathbf{e_i}$ for all $i$. By Theorem 6.77, there is an $\mathbf{a}$ such that $\mathbf{a}' = \mathbf{d}$. Then $\mathbf{a}'' = \mathbf{g}$, so

$$\mathbf{a}'' \not\geq \mathbf{0}'''.$$

If $\mathbf{e_i} = \mathbf{c_i}$ then $(\mathbf{a}' \vee \mathbf{c_i})' = (\mathbf{d} \vee \mathbf{e_i})' = \mathbf{g}' \geq \mathbf{0}'''$. If $\mathbf{e_i} \neq \mathbf{c_i}$ then $(\mathbf{a}' \vee \mathbf{c_i})' \geq \mathbf{a}'' \vee \mathbf{c_i} = \mathbf{g} \vee \mathbf{c_i} \geq \mathbf{d_i} \vee \mathbf{c_i} \geq \mathbf{0}'''$. Thus

$$(\mathbf{a}' \vee \mathbf{c_i})' \geq \mathbf{0}'''$$

for all $i$. These two displayed properties of $\mathbf{a}$ are all that we will use below.

Let $\mathbf{p}$ be a degree that is PA over $\mathbf{a}''$ and hyperimmune-free over $\mathbf{a}''$. By the relativized form of Theorem 6.13, there is an $\mathbf{a}$-computable 2-coloring $d$ of $[\mathbb{N}]^3$ such that for any infinite homogeneous set $H$ for $d$, we have $\deg(H) \vee \mathbf{a} \geq \mathbf{a}'$. We can combine the colorings $c$ and $d$ into an $\mathbf{a}$-computable 4-coloring $e$ of $[\mathbb{N}]^3$ such that any homogeneous set for $e$ is also homogeneous for both $c$ and $d$. (Just let $e(s) = 2c(s) + d(s)$.) By the relativized form of Exercise 6.79, there is an infinite homogeneous set $H$ for $e$ such that $(\deg(H) \vee \mathbf{a})' \leq \mathbf{p}$. Let $\mathbf{b} = \deg(H) \vee \mathbf{a}$. Then $\mathbf{b} \geq \mathbf{a}'$ and $\mathbf{b}'$ is hyperimmune-free over $\mathbf{a}''$.

If $\mathbf{b}' \geq \mathbf{0}'''$ then, since we also have $\mathbf{b}' \geq \mathbf{a}''$, it follows that $\mathbf{a}'' \vee \mathbf{0}'''$ is hyperimmune-free over $\mathbf{a}''$. But $\mathbf{a}'' \vee \mathbf{0}'''$ is c.e. relative to $\mathbf{a}''$, and as mentioned following Theorem 3.13, the only hyperimmune-free c.e. (or even $\Delta_2^0$) sets are the computable ones. Relativizing this fact we conclude that $\mathbf{a}'' \geq \mathbf{0}'''$, which is not the case. Thus $\mathbf{b}' \not\geq \mathbf{0}'''$, and hence $H' \not\geq_T \emptyset'''$.

Now suppose that $\mathbf{c_i} \leq \mathbf{b}$. Then $\mathbf{b} \geq \mathbf{b} \vee \mathbf{c_i} \geq \mathbf{a}' \vee \mathbf{c_i}$, so $\mathbf{b}' \geq (\mathbf{a}' \vee \mathbf{c_i})' \geq \mathbf{0}'''$, which is not the case. Thus $\mathbf{c_i} \not\leq \mathbf{b}$ for all $i$, and hence $C_i \not\leq_T H$ for all $i$. $\qquad \square$

---

$\boxed{55}$ **Exercise 6.80** (Cholak, Jockusch, and Slaman [20]). Complete the proof of Theorem 6.75 by proving the $n = 2$ case (with a similar argument to the $n = 3$ case above) and the inductive case.

## 6.8 Ramsey's Theorem and first order axioms

We have seen that $\mathrm{RCA}_0 \vdash \mathrm{RT}_k^1$ for all $k \in \omega$. However, proving this fact seems to require either an external induction (i.e., one carried out outside $\mathrm{RCA}_0$), or a proof scheme yielding a separate proof for each $k \in \omega$, which again requires moving outside $\mathrm{RCA}_0$ to witness that the scheme works for all $k \in \omega$. It is thus interesting to ask whether such external methods are indeed necessary. In other words, can $\mathrm{RT}_{<\infty}^1$ be proved in $\mathrm{RCA}_0$? We may view $\mathrm{RT}^1_{<\infty}$ as saying that any partition of $\mathbb{N}$ into finitely many parts contains an infinite part. It might seem strange that such a statement could fail to be provable in $\mathrm{RCA}_0$, but we have already discussed this phenomenon in Section 4.1. To remind ourselves of the issue: As noted above, $\mathrm{RT}^1_{<\infty}$ is true in every $\omega$-model of $\mathrm{RCA}_0$, or indeed, any $\omega$-model at all. However, not every model of $\mathrm{RCA}_0$ is an $\omega$-model. In a non-$\omega$-model $\mathcal{M}$, whose first order part is a nonstandard model $N$, the principle $\mathrm{RT}^1_{<\infty}$ implies $\mathrm{RT}^1_k$ for *all* $k$ in $N$, including the nonstandard elements. For a nonstandard $k$, there could well be a function $c : N \to k$ in $\mathcal{M}$ such that $c^{-1}(i)$ is bounded in $N$ for each $i < k$. Indeed, $\mathrm{RT}^1_{<\infty}$ is equivalent to the weakest of the hierarchy of first order statements not provable in $\mathrm{RCA}_0$ introduced in Section 4.6.

**Theorem 6.81** (Hirst [90]). $\mathrm{RT}^1_{<\infty}$ *is equivalent to* $\mathrm{B}\Sigma^0_2$ *over* $\mathrm{RCA}_0$.

*Proof.* By Exercise 4.31, it is enough to prove the equivalence of $\mathrm{RT}^1_{<\infty}$ with $\mathrm{B}\Pi^0_1$.

First assume $\mathrm{B}\Pi^0_1$. Let $c : \mathbb{N} \to k$. Assume for a contradiction that $c$ has no infinite homogeneous set. Then for each $i < k$, the set $c^{-1}(i)$ is bounded. That is, $\forall i < k \, \exists n \, \forall x > n \, [c(x) \neq i]$. By $\mathrm{B}\Pi^0_1$, there is a $b$ such that $\forall i < k \, \exists n < b \, \forall x > n \, [c(x) \neq i]$. But then $c(b) \neq i$ for all $i < k$, which is a contradiction.

Now assume $\mathrm{RT}^1_{<\infty}$. Let $\varphi$ be a bounded quantifier formula and let $w$ be such that $\forall x < w \, \exists y \, \forall z \, \varphi(x, y, z)$. Let $c(n)$ be the least $b < n$ such that $\forall x < w \, \exists y < b \, \forall z < n \, \varphi(x, y, z)$, if such a $b$ exists, and let $c(n) = n$ otherwise. Suppose that we can show that the range of $c$ is bounded. Then $c$ has an infinite homogeneous set $H$. Let $b$ be such that $H$ is homogeneous to $b$ and fix $x < w$. If $n > b$ then $\exists y < b \, \forall z < n \, \varphi(x, y, z)$. Let $d(n)$ be the least such $y$. (For $n \leqslant b$, let $d(n) = 0$.) Then $d$ has an infinite homogeneous set $I$. Let $y$ be such that $I$ is homogeneous to $y$. Then there are infinitely many $n$ such that $\forall z < n \, \varphi(x, y, z)$, so in fact $\forall z \, \varphi(x, y, z)$. Thus $\forall x < w \, \exists y < b \, \forall z \, \varphi(x, y, z)$.

So we are left with showing that the range of $c$ is bounded. Assume for a contradiction that it is not, and let $n_0 < n_1 < \cdots$ be such that $c(n_0) < c(n_1) < \cdots$. Then for each $i$ we have $\exists x < w \, \forall y < c(n_i) \, \exists z < n_{i+1} \, \neg\varphi(x, y, z)$, as otherwise we would have $c(n_{i+1}) \leqslant c(n_i)$. Let $e(i)$ be the least such $x$. Then $e$ has an infinite homogeneous set $J$. Let $x$ be such that $J$ is homogeneous to $x$. For each $y$, there is an $i \in J$ such that $c(n_i) > y$, whence $\exists z \, \neg\varphi(x, y, z)$. But then $\neg\exists y \, \forall z \, \varphi(x, y, z)$, contrary to hypothesis. $\square$

As we will see in Sections 7.2 and 7.3, the above result implies that $RT^1_{<\infty}$ is provable in neither $WKL_0$ nor $RCA_0 + COH$. It is of course provable in $ACA_0$, and hence in $RCA_0 + RT^3_2$, and is a special case of $RT^2_{<\infty}$, so it is natural to ask whether it is also provable in $RT^2_2$, or even $SRT^2_2$, over $RCA_0$. Hirst [90] showed that $RT^1_{<\infty}$ is provable in $RCA_0 + RT^2_2$. This result was later strengthened as follows.

**Theorem 6.82** (Cholak, Jockusch, and Slaman [20]). $RCA_0 + SRT^2_2 \vdash RT^1_{<\infty}$.

*Proof.* Let $c : \mathbb{N} \to k$. Let $d : [\mathbb{N}]^2 \to 2$ be defined by letting $d(x,y) = 0$ if $c(x) \neq c(y)$ and $d(x,y) = 1$ otherwise. Suppose there is an $x$ such that $d(x,y) = 1$ for infinitely many $y$. Then the set of all such $y$ is an infinite homogeneous set for $c$. Otherwise, $d$ is stable, so it has an infinite homogeneous set $H$. The set $H$ cannot be homogeneous to 0, as then the image of $H$ under $c$ would be infinite. So $H$ is homogeneous to 1, and hence is also homogeneous for $c$. $\square$

One of the major lines of research in reverse mathematics is pinpointing the arithmetic (i.e., first order) consequences of second order principles. We mentioned in Section 4.6 that the arithmetic sentences provable in $RCA_0$ or $WKL_0$ are exactly the ones provable in $\Sigma^0_1$-PA, and we will see in Section 7.3 that the same is true of $RCA_0 + COH$. Theorem 6.82 shows that $SRT^2_2$ is more powerful in this sense, as it implies $B\Sigma^0_2$ over $RCA_0$. On the other hand, we have the following theorem, which is an example of a conservativity result; such results will be the topic of the next chapter.

**Theorem 6.83** (Cholak, Jockusch, and Slaman [20]). *Every $\Pi^1_1$ sentence (and hence in particular every arithmetic sentence) provable in $RCA_0 + I\Sigma^0_2 + RT^2_2$ is provable in $RCA_0 + I\Sigma^0_2$.*

This theorem can be seen as a strong form of Seetapun's Theorem that $RCA_0 + RT^2_2$ does not imply $ACA_0$ over $RCA_0$, as it shows that $RCA_0 + RT^2_2$ does not even imply PA, which, as mentioned in Section 4.6 and proved in Section 7.1 below, is the first order part of $ACA_0$.

For a principle $P$, let $(P)^1$ be the first order part of $RCA_0 + P$, i.e., the arithmetic sentences provable in $RCA_0 + P$. We have seen that both $(SRT^2_2)^1$ and $(RT^2_2)^1$ are somewhere between $(B\Sigma^0_2)^1$ and $(I\Sigma^0_2)^1$. The model of $SRT^2_2$ built by Chong, Slaman, and Yang [24] in their proof of Theorem 6.36 is also built to not be a model of $I\Sigma^0_2$, which gives us the only additional information we currently have on $(SRT^2_2)^1$.

**Theorem 6.84** (Chong, Slaman, and Yang [24]). $\mathrm{SRT}_2^2$ *does not imply* $\mathrm{I}\Sigma_2^0$ *over* $\mathrm{RCA}_0$.

Recently, Chong, Slaman, and Yang [personal communication] have extended this result to prove the following theorem.

**Theorem 6.85** (Chong, Slaman, and Yang [in preparation]). $\mathrm{RT}_2^2$ *does not imply* $\mathrm{I}\Sigma_2^0$ *over* $\mathrm{RCA}_0$.

Thus the current state of our knowledge on the first order consequences of $(\mathrm{S})\mathrm{RT}_2^2$ is as follows:

$$(\mathrm{B}\Sigma_2^0)^1 \subseteq (\mathrm{SRT}_2^2)^1 \subsetneq (\mathrm{I}\Sigma_2^0)^1$$
$$(\mathrm{B}\Sigma_2^0)^1 \subseteq (\mathrm{RT}_2^2)^1 \subsetneq (\mathrm{I}\Sigma_2^0)^1.$$

▶ **Open Question 6.86.** Is there any formula in $(\mathrm{RT}_2^2)^1$ (or $(\mathrm{SRT}_2^2)^1$) not provable in $\mathrm{RCA}_0 + \mathrm{B}\Sigma_2^0$? Does $(\mathrm{SRT}_2^2)^1 = (\mathrm{RT}_2^2)^1$?

Chong [personal communication] has pointed out that the principle BME defined in Chong, Slaman, and Yang [24] is a good candidate for a first order principle provable from $\mathrm{SRT}_2^2$ but not from $\mathrm{B}\Sigma_2^0$.

In the absence of an exact characterization of the first order consequences of a principle such as $\mathrm{RT}_2^2$, it is interesting to consider its power to prove particular mathematical statements at that level. In the case of $\mathrm{RT}_2^2$, the appropriate level is that of principles provable from $\mathrm{I}\Sigma_2^0$ but not from $\mathrm{B}\Sigma_2^0$. Montalbán [146] asks in particular whether $\mathrm{RCA}_0 + \mathrm{RT}_2^2$ proves that the Ackermann function is total, and whether this system proves that the length-lexicographic order on $\mathbb{N}^{<\mathbb{N}}$ is a well-order (which is a way of saying that the ordinal $\omega^\omega$ is well-ordered).

Theorem 6.83 allows us to separate $\mathrm{RT}_2^2$ from $\mathrm{RT}_{<\infty}^2$, which, of course, no result relying solely on computability theory and $\omega$-models could do. Before doing so, recall the principle $\mathrm{SRT}_{<\infty}^2$ defined in Section 6.4.

$\boxed{56}$ **Exercise 6.87** (Cholak, Jockusch, and Slaman [20]). Show that $\mathrm{RT}_{<\infty}^2$ is equivalent to $\mathrm{COH} + \mathrm{SRT}_{<\infty}^2$ over $\mathrm{RCA}_0$.

The proof of Theorem 6.36 does not seem to be immediately adaptable to the case of arbitrarily many colors, so the following question is still open.

▶ **Open Question 6.88.** Does $\mathrm{SRT}_{<\infty}^2$ imply $\mathrm{RT}_{<\infty}^2$ over $\mathrm{RCA}_0$?

We can separate $\mathrm{RT}_2^2$ and $\mathrm{SRT}_2^2$ on the one hand from $\mathrm{RT}_{<\infty}^2$ and $\mathrm{SRT}_{<\infty}^2$ on the other by combining Theorem 6.83 with the following result (and Theorem 4.32).

**Theorem 6.89** (Cholak, Jockusch, and Slaman [20]). $\mathrm{RCA}_0 + \mathrm{SRT}^2_{<\infty} \vdash$ $\mathrm{B}\Sigma^0_3$.

*Proof.* By Exercise 4.31, it is enough to show that $\mathrm{SRT}^2_{<\infty}$ implies $\mathrm{B}\Pi^0_2$ over $\mathrm{RCA}_0$. Since $\mathrm{SRT}^2_2$ implies $\mathrm{B}\Sigma^0_2$ over $\mathrm{RCA}_0$, so does $\mathrm{SRT}^2_{<\infty}$, so we are free to use $\mathrm{B}\Sigma^0_2$ in our proof.

Suppose that $\mathrm{B}\Pi^0_2$ fails. That is, there are a $k$ and a bounded quantifier formula $\varphi$ such that for each $i < k$ we have $\exists x \, \forall y \, \exists z \, \varphi(i, x, y, z)$ but for each $b$ there is an $i < k$ with

$$\forall x < b \, \exists y \, \forall z \, \neg \varphi(i, x, y, z). \qquad (6.4)$$

We use these facts to define a stable $k$-coloring $c$ of $[\mathbb{N}]^2$ with no infinite homogeneous set. The basic idea is to ensure that $\lim_s c(b, s)$ is an $i$ as in (6.4). Then an infinite homogeneous set gives us infinitely many $b$ such that (6.4) holds for the same $i$, which cannot happen.

For $i < k$, let $r(i, b, s)$ be the largest $r$ such that

$$\exists x < b \, \forall y < r \, \exists z < s \, \varphi(i, x, y, z),$$

and let $r(b, s) = \min_{i<k} r(i, b, s)$. For $s > b$, let $c(b, s)$ be the least $i < k$ such that $r(i, b, s) = r(b, s)$.

We first show that $c$ is stable. Fix $b$, and let $i$ be as in (6.4). By $\mathrm{B}\Pi^0_1$, there is a $d$ such that $\forall x < b \, \exists y < d \, \forall z \, \neg \varphi(i, x, y, z)$. Then $r(b, s) \leqslant r(i, b, s) < d$ for all $s$. Since $r(b, s)$ is nondecreasing with $s$, it reaches a limit. There is a least $j$ such that $r(j, b, s)$ never exceeds this limit. Then $c(b, s) = j$ for all sufficiently large $s$.

Now assume for a contradiction that $c$ has an infinite homogeneous set $H$, and let $i < k$ be such that $H$ is homogeneous to $i$. Let $x$ be such that $\forall y \, \exists z \, \varphi(i, x, y, z)$. Let $b > x$ be an element of $H$. For each $r$ there is an $s$ such that $\forall y < r \, \exists z < s \, \varphi(i, x, y, z)$, so $r(i, b, s)$ goes to infinity with $s$. But we have shown that $r(b, s)$ reaches a limit, so we cannot have $\lim_s c(b, s) = i$, contradicting the choice of $b$. $\qquad \square$

**Corollary 6.90** (Cholak, Jockusch, and Slaman [20]). $\mathrm{RCA}_0 + \mathrm{RT}^2_2 \nvdash \mathrm{SRT}^2_{<\infty}$.

We have the following analog to Theorem 6.83.

**Theorem 6.91** (Cholak, Jockusch, and Slaman [20]). *Every $\Pi^1_1$ sentence provable in $\mathrm{RCA}_0 + \mathrm{I}\Sigma^0_3 + \mathrm{RT}^2_{<\infty}$ is provable in $\mathrm{RCA}_0 + \mathrm{I}\Sigma^0_3$.*

However, we do not have analogs to Theorems 6.84 and 6.85, so all we know about the first order parts of $(S)RT^2_{<\infty}$ is the following:

$$(B\Sigma^0_3)^1 \subseteq (SRT^2_{<\infty})^1 \subseteq (I\Sigma^0_3)^1$$
$$(B\Sigma^0_3)^1 \subseteq (RT^2_{<\infty})^1 \subseteq (I\Sigma^0_3)^1.$$

▶ **Open Question 6.92.** Does $RT^2_{<\infty}$ (or $SRT^2_{<\infty}$) imply $I\Sigma^0_3$ over $RCA_0$? Is there any formula in $(RT^2_{<\infty})^1$ (or $(SRT^2_{<\infty})^1$) not provable in $RCA_0 +$ $B\Sigma^0_3$? Does $(SRT^2_{<\infty})^1 = (RT^2_{<\infty})^1$?

As mentioned above, Theorems 6.83 and 6.91 are conservativity results. Such results play an important role in reverse mathematics. Proving them requires us to find ways to extend arbitrary countable models (rather than just $\omega$-models, as we have done so far), and will be the topic of Chapter 7.

## 6.9   Uniformity

We finish this chapter with a few facts about uniformity. For more on uniform computability theoretic reducibility (which, as mentioned in Section 2.2, is equivalent to a special case of Weihrauch reducibility), see Dorais, Dzhafarov, Hirst, Mileti, and Shafer [38] and Hirschfeldt and Jockusch [85]. Recall that $P \leqslant_u Q$ means that there are Turing functionals $\Phi$ and $\Psi$ such that if $X$ is an instance of $P$ then $\Phi^P$ is an instance of $Q$, and for any solution $Z$ to $\Phi^P$, the set $\Psi^{X \oplus Z}$ is a solution to $X$. We have already mentioned that if $n \geqslant 1$ and $1 < j < k$ then $RT^n_k \not\leqslant_u RT^n_j$, that $KL \leqslant_u RT^3_2$, that $COH \leqslant_u RT^2_2$, that $COH \not\leqslant_u SRT^2_2$, and that $RT^2_2 \not\leqslant_u KL$.

The issues involved in working with $\leqslant_u$ can be a bit subtle. For instance, consider $RT^1_{<\infty}$. One way to formalize this principle is to have an instance be a function $f : \mathbb{N} \to \mathbb{N}$ such that $rng(f)$ is bounded (and a solution to this instance be an infinite set $H$ such that $rng(f \restriction H)$ is a singleton). Another is to have an instance consist of a number $k$ together with a function $f : \mathbb{N} \to k$. It is not very difficult to show that these two notions are different with respect to uniform reducibility. As the latter notion is probably closer to the intent of the principle, let us call it $RT^1_{<\infty}$ (although the results in the following exercise also hold for the former notion).

57 **Exercise 6.93** (Hirschfeldt and Jockusch [85]).
a. Show that $RT^1_{<\infty} \leqslant_u RT^2_2$.
b. Show that $RT^1_{<\infty} \leqslant_u KL$.
c. Show that $RT^1_2 \leqslant_u SRT^2_2$.

**d.** Show that $\mathrm{RT}^1_3 \not\leq_u \mathrm{SRT}^2_2$. [Hint: Suppose $c$ is a stable 2-coloring of pairs and $\Psi$ is a Turing functional such that $\Psi^H$ is total and infinite for every infinite homogeneous set $H$ for $c$. If there is an infinite homogeneous set $H$ for $c$ that is homogeneous to $i$ then there is a finite set $F \subset H$ and an $n$ such that $\Psi^F(n)\!\downarrow = 1$. While we cannot find such an $F$ effectively from $c$, we can make guesses that will eventually be correct.]

**e.** Show that $\mathrm{RT}^1_2 \not\leq_u \mathrm{COH}$.

**f.** Show that $\mathrm{RT}^1_2 \not\leq_u \mathrm{WKL}$. [Hint: Use the fact that if $T$ is a binary tree and $\Psi$ is a Turing functional such that $\Psi^X$ is total and infinite (when thought of as a set) for all $X \in [T]$, then we can $T$-computably find an $n$ such that for each $\sigma$ of length $n$, there is an $m$ for which $\Psi^\sigma(m)\!\downarrow = 1$.]

**g.** Show that $\mathrm{SRT}^2_2 \not\leq_u \mathrm{KL}$. [Hint: One way to proceed is as in the proof of the previous part, but "one jump up". Instead of building a stable coloring of pairs directly, build an $\emptyset'$-computable 2-coloring $c$ of singletons. By the recursion theorem and the limit lemma, we may assume that we know an index for the reduction from $\emptyset'$ to $c$, and hence know an index for a computable stable coloring $d$ of pairs such that $c(x) = \lim_s d(x,s)$. Now use (the proof of) Exercise 3.27.]

**h.** Show that $\mathrm{COH} \leq_u \mathrm{KL}$. [Hint: One way to proceed is, given an instance of COH, to build an $\emptyset'$-computable binary tree whose paths encode infinite cohesive sets and apply (the proof of) Exercise 3.27.]

# Chapter 7

# Preserving Our Power: Conservativity

Sometimes it is possible to prove theorems of the form "$P$ does not imply $Q$ over the base theory $R$" for a whole class of principles $Q$ at once, based purely on their logical form. Such theorems can give us a great deal of information about the power of $R + P$ relative to the power of $R$ alone.

**Definition 7.1.** Let $T_0 \subseteq T_1$ be theories in languages $\mathcal{L}_0 \subseteq \mathcal{L}_1$, respectively. Let $\Gamma$ be a collection of $\mathcal{L}_0$-sentences. We say that $T_1$ is $\Gamma$-*conservative* (or *conservative for $\Gamma$ sentences*) *over* $T_0$ if for every $\sigma \in \Gamma$, we have $T_1 \vdash \sigma$ iff $T_0 \vdash \sigma$. If $\Gamma$ is the collection of all $\mathcal{L}_0$-sentences, then we say that $T_1$ is a *conservative extension* of $T_0$.

Thus Theorems 6.83 and 6.91 say that $\mathrm{RCA}_0 + \mathrm{I}\Sigma_2^0 + \mathrm{RT}_2^2$ is $\Pi_1^1$-conservative over $\mathrm{RCA}_0 + \mathrm{I}\Sigma_2^0$, while $\mathrm{RCA}_0 + \mathrm{I}\Sigma_3^0 + \mathrm{RT}_{<\infty}^2$ is $\Pi_1^1$-conservative over $\mathrm{RCA}_0 + \mathrm{I}\Sigma_3^0$. (When discussing the conservativity of $\mathrm{RCA}_0 + P + Q$ over $\mathrm{RCA}_0 + P$, we sometimes say more informally that $Q$ is conservative over $P$. For example, we say that $\mathrm{RT}_2^2$ is $\Pi_1^1$-conservative over $\mathrm{I}\Sigma_2^0$.)

In the context of first order logic (and countable languages), we can prove that $T_1$ is $\Gamma$-conservative over $T_0$ by showing that every countable model $\mathcal{M}$ of $T_0$ is a submodel of (the restriction to $\mathcal{L}_0$ of) some model $\mathcal{N}$ of $T_1$ such that any sentence in $\Gamma$ that fails to hold in $\mathcal{M}$ also fails to hold in $\mathcal{N}$: Suppose that is the case. If $\varphi \in \Gamma$ and $T_0 \nvdash \varphi$, then by the downward Löwenheim-Skolem-Tarski Theorem, there is a countable model $\mathcal{M}$ of $T_0 + \neg\varphi$, and the corresponding model $\mathcal{N}$ is then a model of $T_1 + \neg\varphi$, so $T_1 \nvdash \varphi$. (The fact that we can restrict attention to countable models $\mathcal{M}$ is useful, allowing us for example to build $\mathcal{N}$ using forcing constructions, as we will see below. For more on basic model theoretic results such as the LST Theorem, see for instance Enderton [53].)

It is important to recall here that what we call "second order arithmetic"

125

is actually a first order theory in a two-sorted language, so the above procedure applies to our subsystems of second order arithmetic. A structure in the language of second order arithmetic is a first order structure; when we say that such a structure $(N, S, +_N, \cdot_N, 0_N, 1_N, \leqslant_N)$ is countable, we mean that both $N$ and $S$ are countable.

In the $\Gamma$-conservativity results we will discuss, $\Gamma$ will contain at least all arithmetic sentences, which leads us to the following concept.

**Definition 7.2.** We say that $\mathcal{M} = (M, S_M, +_M, \cdot_M, 0_M, 1_M, \leqslant_M)$ is an $\omega$-*submodel* of $\mathcal{N} = (N, S_N, +_N, \cdot_N, 0_N, 1_N, \leqslant_N)$, and write $\mathcal{M} \subseteq_\omega \mathcal{N}$, if the two models have the same first order part (i.e., $(M, +_M, \cdot_M, 0_M, 1_M, \leqslant_M) = (N, +_N, \cdot_N, 0_N, 1_N, \leqslant_N))$ and $S_M \subseteq S_N$. In this case, we also say that $\mathcal{N}$ is an $\omega$-*extension* of $\mathcal{N}$.

Note that saying that $\mathcal{M}$ is an $\omega$-submodel of $\mathcal{N}$ does not imply that $\mathcal{M}$ and $\mathcal{N}$ are $\omega$-models. (A model $\mathcal{M}$ is an $\omega$-model iff it is an $\omega$-submodel of $(\omega, \mathcal{P}(\omega), +, \cdot, 0, 1, \leqslant)$.)

If $T_0 \subseteq T_1$ are two theories in the language of second order arithmetic and every countable model of $T_0$ is an $\omega$-submodel of a model of $T_1$, then by the argument given above, $T_1$ is conservative for arithmetic sentences over $T_0$. Indeed, even more is true in this case. If $\mathcal{M} \subseteq_\omega \mathcal{N}$ then it is easy to see that any $\Pi_1^1$ statement that fails to hold in $\mathcal{M}$ also fails to hold in $\mathcal{N}$, so proofs of conservativity for arithmetic sentences using $\omega$-submodels as above actually yield proofs of $\Pi_1^1$-conservativity.

In this chapter we will discuss several conservativity results. We will begin by showing that $\mathrm{ACA}_0$ and $\mathrm{RCA}_0$ are conservative extensions of PA and $\Sigma_1^0$-PA, respectively, as mentioned in Section 4.6. Then we will show that $\mathrm{WKL}_0$ is $\Pi_1^1$-conservative over $\mathrm{RCA}_0$, and hence is also a conservative extension of $\Sigma_1^0$-PA. Next, we will show that $\mathrm{RCA}_0 + \mathrm{COH}$ is conservative over $\mathrm{RCA}_0$ for a collection of sentences properly containing the $\Pi_1^1$ sentences, which will not only show that $\mathrm{RCA}_0 + \mathrm{COH}$ is yet another conservative extension of $\Sigma_1^0$-PA, but also give us a uniform proof of nonimplication between COH and a large class of principles (including $(\mathrm{S})\mathrm{RT}_2^2$ and $\mathrm{WKL}_0$) over $\mathrm{RCA}_0$.

The conservativity proofs for $\mathrm{WKL}_0$ and COH will use forcing constructions. We have already seen, in the proofs of Corollary 6.51 and of Seetapun's Theorem, one way in which forcing can be used to prove nonimplication results. We start with a countable model $\mathcal{M}$ of $\mathrm{RCA}_0$ in which some particular instance $I$ of a principle $P$ expressed by a sentence of the form

$$\forall X \, [\theta(X) \to \exists Y \, \psi(X, Y)] \tag{7.1}$$

with $\theta$ and $\psi$ arithmetic fails. We then use a forcing construction to add a solution to some instance of another principle $Q$ of the form (7.1) without adding a solution to $I$. We iterate this process to obtain in the end a model $\mathcal{N}$ of $\mathrm{RCA}_0 + Q$ in which $I$ still has no solution, and hence $P$ fails, thus showing that $Q$ does not imply $P$ over $\mathrm{RCA}_0$. (The fact that $\theta$ and $\psi$ are arithmetic ensures that these formulas do not change meaning as we expand $\mathcal{M}$, since we never change the first order part of our models.) In the cases we have considered, and many others, we can take $\mathcal{M}$ to be an $\omega$-model. When proving conservativity results, we can still use a version of this technique, but we must work over arbitrary countable models.

That is, we start with an arbitrary countable model $\mathcal{M}$ of $T_0$ and use a forcing construction to obtain a generic object $G$ and a new model $\mathcal{M}[G]$ that is still a model of $T_0$, is not a model of $\varphi$ for any $\varphi \in \Gamma$ that does not hold in $\mathcal{M}$, and makes some progress towards being a model of $T_1$ (if $T_1 = T_0 + P$ for a principle $P$ of the form (7.1), $\mathcal{M}[G]$ may for instance contain a solution to a given instance of $P$ in $\mathcal{M}$). The model $\mathcal{M}[G]$ is obtained by adding $G$ to $\mathcal{M}$ and "closing off" to obtain a model of $T_0$. For instance, if $T_0$ is $\mathrm{RCA}_0$ and $S$ is the second order part of $\mathcal{M}$, then we need to add all sets that are $\Delta_1^0$-definable with parameters in $S \cup \{G\}$. Finally, we iterate this process to obtain a model $\mathcal{N} = \mathcal{M}[G_0][G_1] \cdots$ of $T_1$ such that if $\varphi \in \Gamma$ fails to hold in $\mathcal{M}$, then $\varphi$ fails to hold in $\mathcal{N}$. (See Section 7.2 for more details.)

Of course, we have to be careful to remember that our notions of forcing are defined within our models, which are not in general $\omega$-models, so notions like "finite" and "infinite" in these definitions have to be taken in the sense of the given model (see Section 4.2). For a structure $\mathcal{M}$ in the language of second order arithmetic, we denote the domain of the first order part of $\mathcal{M}$ by $M$, the second order part of $\mathcal{M}$ by $S_M$, and the order relation on the first order part of $\mathcal{M}$ by $\leqslant_M$. When we say that a formula has parameters from $\mathcal{M}$, we mean that all its parameters are in $M \cup S_M$.

## 7.1 Conservativity over first order systems

In this section, we prove the facts about the first order parts of $\mathrm{ACA}_0$ and $\mathrm{RCA}_0$ mentioned in Sections 4.1 and 4.3. We will use the following notions.

**Definition 7.3.** Let $\mathcal{M}$ be a structure in the language of second order arithmetic. A set $X \subseteq M$ is *arithmetically definable over* $\mathcal{M}$ if $X = \{n \in M : \mathcal{M} \vDash \varphi(n)\}$ for some arithmetic formula $\varphi(x)$ with parameters from $\mathcal{M}$ and no free variables other than $x$.

A set $X \subseteq M$ is $\Delta_1^0$-*definable over* $\mathcal{M}$ if $X = \{n \in M : \mathcal{M} \vDash \varphi(n)\} = \{n \in M : \mathcal{M} \vDash \psi(n)\}$ for some $\Sigma_1^0$ formula $\varphi(x)$ and some $\Pi_1^0$ formula $\psi(x)$, both with parameters from $\mathcal{M}$ and no free variables other than $x$.

We begin with $\mathrm{ACA}_0$. Recall that $P_0^-$ is the collection of basic first order axioms from Definition 4.1.

**Theorem 7.4** (Friedman [60]). *If $\mathcal{M}$ satisfies $P_0^-$ and arithmetic induction, then it is an $\omega$-submodel of some model of $\mathrm{ACA}_0$.*

*Proof.* Let $S_N$ be the set of $X \subseteq M$ that are arithmetically definable over $\mathcal{M}$. Let $\mathcal{N}$ be the $\omega$-extension of $\mathcal{M}$ with second order part $S_N$. Then $\mathcal{N}$ satisfies $P_0^-$. Furthermore, if $X \in S_N$ and $\varphi$ is an arithmetic formula defining $X$ over $\mathcal{M}$, then $\mathcal{M}$ satisfies induction for $\varphi$, and hence $\mathcal{N}$ satisfies the instance of set induction corresponding to $X$ (see Definition 4.1). Since $X$ is arbitrary, $\mathcal{N}$ satisfies set induction. Thus we are left with showing that $\mathcal{N}$ satisfies arithmetic comprehension.

Let $\psi(x)$ be an arithmetic formula with parameters from $\mathcal{N}$ and no free variables other than $x$. Let $Y_0, \ldots, Y_{l-1}$ be the second order parameters in $\psi$. Let $\varphi_0, \ldots, \varphi_{l-1}$ be arithmetic formulas defining $Y_0, \ldots, Y_{l-1}$, respectively, over $\mathcal{M}$. We can replace all of the occurrences of atomic formulas $t \in Y_i$ in $\psi$ by $\varphi_i(t)$, to obtain an arithmetic formula $\theta$ with parameters in $\mathcal{M}$. Then $\mathcal{N}$ satisfies both $\exists X \, \forall n \, [n \in X \leftrightarrow \theta(n)]$ (because $\{n \in M : \mathcal{M} \vDash \theta(n)\} \in S_N$) and $\forall n \, [\psi(n) \leftrightarrow \theta(n)]$, so it also satisfies $\exists X \, \forall n \, [n \in X \leftrightarrow \psi(n)]$. $\qquad\square$

Let $\mathcal{M}$ be a model of PA. Although $\mathcal{M}$ is a structure in the language of first order arithmetic, we can think of it as a structure in the language of second order arithmetic with empty second order part. Then $\mathcal{M}$ satisfies $P_0^-$ and arithmetic induction, so by the above theorem, it is an $\omega$-submodel of some model of $\mathrm{ACA}_0$, which as discussed above, gives us the following result.

**Corollary 7.5** (Friedman [60]). $\mathrm{ACA}_0$ *is a conservative extension of* PA.

We now turn to $\mathrm{RCA}_0$.

**Theorem 7.6** (Friedman [60]). *If $\mathcal{M}$ satisfies $P_0^- + \text{I}\Sigma_1^0$, then it is an $\omega$-submodel of some model of $\text{RCA}_0$.*

*Proof.* Let $S_N$ be the set of $X \subseteq M$ that are $\Delta_1^0$-definable over $\mathcal{M}$. Let $\mathcal{N}$ be the model with the same first order part as $\mathcal{M}$ and second order part $S_N$. Then $\mathcal{M}$ is an $\omega$-submodel of $\mathcal{N}$, and hence $\mathcal{N}$ satisfies $P_0^-$. We will show that if $\varphi$ is a $\Sigma_1^0$ formula with parameters from $\mathcal{N}$ and no free set variables, then there is a $\Sigma_1^0$ formula $\psi$ with parameters from $\mathcal{M}$ and the same free variables as $\varphi$ such that $\mathcal{N} \vDash \varphi \leftrightarrow \psi$. It then follows easily from the fact that $\mathcal{M}$ satisfies $\text{I}\Sigma_1^0$ and $\Delta_1^0$-comprehension that so does $\mathcal{N}$.

Writing $\varphi$ as $\exists x\, \theta$, where $\theta$ is a bounded quantifier formula, it is enough to show that there is a $\Sigma_1^0$ formula $\exists y\, \eta$ with parameters from $\mathcal{M}$ and the same free variables as $\theta$ such that $\mathcal{N} \vDash \theta \leftrightarrow \exists y\, \eta$, as then we can take $\psi$ to be $\exists u\, \exists x < u\, \exists y < u\, \eta$, where $u$ is not free in $\varphi$. We proceed by induction on the structure of $\theta$. To do so, we need a stronger induction hypothesis, namely that there are a $\Sigma_1^0$ formula $\exists y\, \theta_\Sigma$ and a $\Pi_1^0$ formula $\forall y\, \theta_\Pi$, both with parameters from $\mathcal{M}$ and the same free variables as $\theta$, such that $\mathcal{N} \vDash \theta \leftrightarrow \exists y\, \theta_\Sigma$ and $\mathcal{N} \vDash \theta \leftrightarrow \forall y\, \theta_\Pi$. Here $y$ is a variable not occurring in $\theta$. We make the common assumption that the only connectives in our language are negation and conjunction, and the only bounded quantifier is the universal one, since this assumption does not diminish the expressive power of the language. We denote by $\nu[y/z]$ the result of substituting $z$ in for $y$ in $\nu$.

We begin with the atomic case. If $\theta$ is of the form $t_1 = t_2$ or $t_1 < t_2$ then let $\theta_\Sigma \equiv \theta_\Pi \equiv \theta$. If $\theta$ is of the form $t \in X$, then $X \in S_N$ (since $\theta$ has no free set variables), so $X$ is $\Delta_1^0$-definable over $\mathcal{M}$. Thus there are bounded quantifier formulas $\nu_\Sigma$ and $\nu_\Pi$ such that

$$X = \{n \in M : \mathcal{M} \vDash \exists y\, \nu_\Sigma(n)\} = \{n \in M : \mathcal{M} \vDash \forall y\, \nu_\Pi(n)\}.$$

Let $\theta_\Sigma \equiv \nu_\Sigma(t)$ and $\theta_\Pi \equiv \nu_\Pi(t)$.

We now handle the inductive cases. If $\theta \equiv \neg\theta'$, then let $\theta_\Sigma \equiv \neg\theta'_\Pi$ and $\theta_\Pi \equiv \neg\theta'_\Sigma$. If $\theta \equiv \theta' \wedge \theta''$ then let

$$\theta_\Sigma \equiv \exists z < y\, \theta'_\Sigma[y/z] \wedge \exists z < y\, \theta''_\Sigma[y/z]$$

and $\theta_\Pi \equiv \theta'_\Pi \wedge \theta''_\Pi$. Finally, if $\theta \equiv \forall w < t\, \theta'$ then let

$$\theta_\Sigma \equiv \forall w < t\, \exists z < y\, \theta'_\Sigma[y/z]$$

and $\theta_\Pi \equiv \forall w < t\, \theta'_\Pi$.

It is easy to check that this definition has the desired properties. In the last inductive case, the equivalence of $\theta_\Sigma$ with $\theta$ follows from the fact that, by Exercise 4.8, $\text{B}\Sigma_1^0$ holds in $\mathcal{M}$. $\qquad\square$

As before, we have the following consequence.

**Corollary 7.7** (Friedman [60]). $RCA_0$ *is a conservative extension of* $\Sigma_1^0$-*PA*.

Notice that the proof of Theorem 7.6 shows that (in the context of structures whose first order parts satisfy $P_0^-$) if we want to extend a given structure to a model $\mathcal{N}$ of $RCA_0$, all we need to do is extend it to a model $\mathcal{M}$ of $I\Sigma_1^0$, and then define $\mathcal{N}$ from $\mathcal{M}$ as in that proof. This fact will be useful in the proof of Proposition 7.10 below.

## 7.2  WKL$_0$ and $\Pi_1^1$-conservativity

The following is probably the best known conservativity result in reverse mathematics. Comments on its significance to foundational concerns in the philosophy of mathematics can be found in Simpson [187, 190, 191].

**Theorem 7.8** (Harrington, see Simpson [191]). $WKL_0$ *is* $\Pi_1^1$-*conservative over* $RCA_0$. *Thus,* $WKL_0$ *is a conservative extension of* $\Sigma_1^0$-*PA*.

To prove this theorem, we first give a general framework for proving $\Pi_1^1$-conservativity results over $RCA_0$.

**Definition 7.9.** Let $\mathcal{M}$ be a structure in the language of second order arithmetic and let $G \subseteq M$. We denote by $\mathcal{M} \cup \{G\}$ the $\omega$-extension of $\mathcal{M}$ with second order part $S_M \cup \{G\}$. We denote by $\mathcal{M}[G]$ the $\omega$-extension of $\mathcal{M}$ with second order part consisting of all sets that are $\Delta_1^0$-definable over $\mathcal{M} \cup \{G\}$.

**Proposition 7.10.** *Let* $P \equiv \forall X\,[\theta(X) \rightarrow \exists Y\,\psi(X,Y)]$, *with* $\theta$ *and* $\psi$ *arithmetic. Suppose that for every countable model* $\mathcal{M}$ *of* $RCA_0$ *and every* $X$ *such that* $\mathcal{M} \vDash \theta(X)$, *there is a* $G$ *such that* $\mathcal{M} \cup \{G\} \vDash I\Sigma_1^0 + \psi(X,G)$. *Then* $RCA_0 + P$ *is* $\Pi_1^1$-*conservative over* $RCA_0$.

*Proof.* Let $\mathcal{M}$ be a countable model of $RCA_0$. As discussed above, it is enough to produce an $\omega$-extension $\mathcal{N}$ of $\mathcal{M}$ such that $\mathcal{N} \vDash RCA_0 + P$. We proceed in stages, defining a countable model $\mathcal{M}_n$ at each stage $n \in \omega$. Each $\mathcal{M}_{n+1}$ will be an $\omega$-extension of $\mathcal{M}_n$. Whenever we define $\mathcal{M}_n$, let $X_{n,0}, X_{n,1}, \ldots$ list the elements $X$ of the second order part of $\mathcal{M}_n$ such that $\mathcal{M}_n \vDash \theta(X)$. Note that, since $\theta$ is arithmetic, we also have $\mathcal{M}_{n'} \vDash \theta(X)$ for any $n' > n$.

At stage 0, let $\mathcal{M}_0 = \mathcal{M}$. At stage $n > 0$, let $m < n$ and $k$ be the least pair of numbers for which we have not yet acted. Let $G$ be such that $\mathcal{M}_{n-1} \cup \{G\} \vDash \mathrm{I}\Sigma_1^0 + \psi(X_{m,k}, G)$, and let $\mathcal{M}_n = \mathcal{M}_{n-1}[G]$. Since $\psi$ is arithmetic, $\mathcal{M}_n \vDash \psi(X_{m,k}, G)$, and by the proof of Theorem 7.6, $\mathcal{M}_n \vDash \mathrm{RCA}_0$.

Let $\mathcal{N} = \bigcup_n \mathcal{M}_n$. Then $\mathcal{N} \vDash \mathrm{RCA}_0$. If $X$ is an element of the second order part of $\mathcal{N}$, then $X = X_{m,k}$ for some $m$ and $k$, so there are an $n$ and a $G$ such that $\mathcal{M}_n \vDash \psi(X, G)$, which implies that $\mathcal{N} \vDash \psi(X, G)$. Thus $\mathcal{N} \vDash \mathrm{RCA}_0 + P$, and $\mathcal{M} \subseteq_\omega \mathcal{N}$, as required. $\qquad\square$

$\boxed{58}$ **Exercise 7.11** (Cholak, Jockusch, and Slaman [20]). Use Proposition 7.10 to show that $\mathrm{RCA}_0 + \mathrm{COH}$ is $\Pi_1^1$-conservative over $\mathrm{RCA}_0$. [Hint: Let $G$ be sufficiently generic for the notion of Mathias forcing in $\mathcal{M}$, where conditions are of the form $(F, I)$ with $F$ an $\mathcal{M}$-finite set and $I$ an $\mathcal{M}$-infinite set, and $\max F <_M \min I$. By Exercise 4.8, it is enough to show that $\mathcal{M} \cup \{G\} \vDash \mathrm{I}\Pi_1^0$. To do so, fix a $\Pi_1^0$ formula $\varphi(x, Y)$ with parameters from $\mathcal{M}$ and show that the set of conditions that force $\forall x\, \varphi(x, G) \vee \exists n\, [\neg\varphi(n, G) \wedge \forall x < n\, \varphi(x, G)]$ is dense.]

We now use our general framework to prove Theorem 7.8.

*Proof of Theorem 7.8.* Let $\mathcal{M}$ be a countable model of $\mathrm{RCA}_0$ and let $T_0$ be an infinite binary tree in $\mathcal{M}$. (I.e., let $A \in S_M$ be such that $\mathcal{M}$ satisfies the formula stating that $A$ codes an infinite binary tree, and let $T_0$ be the tree coded by $A$. Such a tree is a set of strings. Of course, here a string is an element of $2^{<\mathbb{N}}$, where $\mathbb{N}$ is the domain $M$ of the first order part of $\mathcal{M}$.) Consider the notion of forcing $P$ consisting of all infinite binary trees in $\mathcal{M}$ ordered by inclusion (i.e., $T$ extends $\widehat{T}$ iff $T \subseteq \widehat{T}$). A subset $S$ of $P$ is $\mathcal{M}$-definable if there is a formula $\varphi(X)$ with parameters from $\mathcal{M}$ and no free variables other than $X$ such that $\mathcal{M} \vDash \varphi(A)$ iff $A$ codes an element of $S$. Let $\mathcal{D}$ be the collection of all $\mathcal{M}$-definable dense subsets of $P$. Then $\mathcal{D}$ is countable, so there is a $\mathcal{D}$-generic filter $\mathcal{F}$ such that $T_0 \in \mathcal{F}$. For each $n \in M$, the set of $T \in P$ such that $T$ contains only one string of length $n$ is in $\mathcal{D}$, so the intersection of $\mathcal{F}$ consists of a single element $G$ of $2^M$, which we think of as an element of $S_M$. Clearly $G$ is a path on $T_0$, so by Proposition 7.10, it is enough to show that $\mathcal{M} \cup \{G\} \vDash \mathrm{I}\Sigma_1^0$.

We will deal with formulas involving strings; translating these formulas to more formal ones involving the codes of these strings is straightforward. The heart of the proof is the following lemma, which shows that for any bounded quantifier formula $\psi(x, \tau)$ with parameters from $\mathcal{M}$, and any $b \in$

$M$, the elements of $P$ that for each $n \leqslant_M b$ force either $\exists z \, \psi(n, G \restriction z)$ or its negation form a dense set. We will then prove a normal form lemma allowing us to put any $\Sigma_1^0$ formula with one free number variable and one free set variable into the form $\exists z \, \psi(n, Y \restriction z)$. The combination of these two lemmas will allow us to show that if $\mathcal{M} \cup \{G\} \vDash \neg \forall x \, \varphi(x, G)$ for a $\Sigma_1^0$ formula $\varphi$, then there is a least $n \in M$ such that $\mathcal{M} \cup \{G\} \vDash \neg \varphi(n, G)$, which shows that $\mathcal{M} \cup \{G\} \vDash \mathrm{I}\Sigma_1^0$.

**Lemma 7.12.** *Let $\psi(x, \tau)$ be a bounded quantifier formula with parameters from $\mathcal{M}$ and no free variables other than $x$ and $\tau$, and let $b \in M$. Let $D$ be the set of $T \in P$ such that, for each $n \leqslant_M b$, we have that $\mathcal{M}$ satisfies either*

1. $\forall \sigma \in T \, \neg \psi(n, \sigma)$ *or*
2. $\exists k \, \forall \sigma \in T \cap 2^k \, \exists j \leqslant k \, \psi(n, \sigma \restriction j)$,

*where $2^k$ is the set of strings of length $k$, and $\sigma \restriction j$ is the initial segment of $\sigma$ of length $j$. Then $D \in \mathcal{D}$.*

*Proof.* It is clear that $D$ is $\mathcal{M}$-definable. We show that it is also dense in $P$. Fix $T \in P$. We define trees $T_\sigma$ for all strings $\sigma$ as follows. For the empty string $\emptyset$, let $T_\emptyset = T$. Given $T_\sigma$, let

$$T_{\sigma 0} = \{ \tau \in T_\sigma : \forall j \leqslant |\tau| \, [\mathcal{M} \vDash \neg \psi(|\sigma|, \tau \restriction j)] \}$$

and $T_{\sigma 1} = T_\sigma$. Note that, since $\psi$ is a bounded quantifier formula, each $T_\sigma$ is in $\mathcal{M}$. Clearly, each $T_\sigma$ extends $T$ in the sense of $P$, so it is enough to show that there is a $\sigma$ such that $T_\sigma \in D$.

Let $I = \{ \sigma \in 2^{b+1} : T_\sigma \text{ is } \mathcal{M}\text{-infinite} \}$. Since $\mathcal{M}$ satisfies bounded $\Sigma_1^0$-comprehension, $I$ is $\mathcal{M}$-finite. It is also nonempty, since it contains the string of $b + 1$ many 1's. Thus $I$ has a lexicographically least element $\sigma$. Let $n \leqslant_M b$. If $\sigma(n) = 0$ then

$$T_\sigma \subseteq T_{\sigma \restriction n+1} = \{ \tau \in T_{\sigma \restriction n} : \forall j \leqslant |\tau| \, [\mathcal{M} \vDash \neg \psi(n, \tau \restriction j)] \} \subseteq \{ \tau : \neg \psi(n, \tau) \},$$

so case 1 in the definition of $D$ holds. If $\sigma(n) = 1$ then $(\sigma \restriction n)0 \notin I$, so $T_{(\sigma \restriction n)0} = \{ \tau \in T_{\sigma \restriction n} : \forall j \leqslant |\tau| \, [\mathcal{M} \vDash \neg \psi(n, \tau \restriction j)] \}$ is $\mathcal{M}$-finite, which implies that case 2 in the definition of $D$ holds. In either case, $T_\sigma \in D$. $\square$

We now need a lemma that establishes a normal form for the $\Sigma_1^0$ formulas we need to consider.

**Lemma 7.13.** *Let $\varphi(Y)$ be a $\Sigma_1^0$ formula with parameters from $\mathcal{M}$ and no free set variables other than $Y$. Then there is a bounded quantifier formula $\psi(\tau)$ with parameters from $\mathcal{M}$, no free set variables, the same free number variables as $\varphi$, and the additional free string variable $\tau$, such that $\mathcal{M} \cup \{G\} \vDash \varphi(G) \leftrightarrow \exists z\, \psi(G \upharpoonright z)$, where $z$ is not among the free variables of $\varphi$.*

*Proof.* Let us say that a formula $\psi$ as in the lemma is a *string formula corresponding to* $\varphi$. Write $\varphi$ as $\exists x\, \theta$, where $\theta$ is a bounded quantifier formula. It is enough to show that there is a string formula $\eta$ corresponding to $\theta$, as then

$$\mathcal{M} \cup \{G\} \vDash \varphi(G) \leftrightarrow \exists x \,\exists z\, \eta(G \upharpoonright z),$$

so we can let $\psi \equiv \exists x < z \,\exists w < z\, \eta(\tau \upharpoonright w)$.

We proceed by induction on the structure of $\theta$. As in the proof of Theorem 7.6, we need a stronger induction hypothesis, namely that there are bounded quantifier formulas $\theta_\Sigma(\tau)$ and $\theta_\Pi(\tau)$, with parameters from $\mathcal{M}$, no free set variables, the same free number variables as $\theta$, and the additional free string variable $\tau$, such that $\mathcal{M} \cup \{G\} \vDash \theta(G) \leftrightarrow \exists z\, \theta_\Sigma(G \upharpoonright z)$ and $\mathcal{M} \cup \{G\} \vDash \theta(G) \leftrightarrow \forall z\, \theta_\Pi(G \upharpoonright z)$. As in the proof of Theorem 7.6, we assume that the only connectives in our language are negation and conjunction, and the only bounded quantifier is the universal one.

Most cases are similar to the corresponding ones in the proof of Theorem 7.6. If $\theta$ is of the form $t_1 = t_2$, $t_1 < t_2$, or $t \in X$ for $X \in S_M$, then let $\theta_\Sigma \equiv \theta_\Pi \equiv \theta$. If $\theta$ is of the form $t \in Y$, then let $\theta_\Sigma \equiv t < z \wedge \tau(t) = 1$ and $\theta_\Pi \equiv t < z \rightarrow \tau(t) = 1$. If $\theta \equiv \neg\theta'$, then let $\theta_\Sigma \equiv \neg\theta_\Pi'$ and $\theta_\Pi \equiv \neg\theta_\Sigma'$. If $\theta \equiv \theta' \wedge \theta''$ then let

$$\theta_\Sigma \equiv \exists w < z\, \theta_\Sigma'(\tau \upharpoonright w) \wedge \exists w < z\, \theta_\Sigma''(\tau \upharpoonright w)$$

and $\theta_\Pi \equiv \theta_\Pi' \wedge \theta_\Pi''$. Finally, if $\theta \equiv \forall w < t\, \theta'$ then let

$$\theta_\Sigma \equiv \forall w < t \,\exists u < z\, \theta_\Sigma'(\tau \upharpoonright u)$$

and $\theta_\Pi \equiv \forall w < t\, \theta_\Pi'$.

In checking that this definition has the desired properties, the only nontrivial case is the $\theta_\Sigma$ part of the last inductive case, which is proved correct as follows. By the inductive hypothesis, we have

$$\mathcal{M} \cup \{G\} \vDash \theta(G) \leftrightarrow \forall w < t \,\exists z\, \theta_\Sigma'(G \upharpoonright z).$$

Clearly

$$\mathcal{M} \cup \{G\} \vDash [\exists z \,\forall w < t \,\exists u < z\, \theta_\Sigma'(G \upharpoonright u)] \rightarrow [\forall w < t \,\exists z\, \theta_\Sigma'(G \upharpoonright z)].$$

We now show that the converse implication also holds in $\mathcal{M} \cup \{G\}$.

Let $\widehat{\theta}'_\Sigma(w, \tau)$ be obtained by replacing the free number variables of $\theta'$ other than $w$ by elements of $M$, let $b \in M$, and suppose that $\mathcal{M} \cup \{G\} \vDash \forall w < b \, \exists z \, \widehat{\theta}'_\Sigma(w, G \upharpoonright z)$. Let $D$ be as in Lemma 7.12 with $\psi = \widehat{\theta}'_\Sigma(w, \tau)$. Then there is a $T \in D$ such that $G$ is a path on $T$. Case 1 in the definition of $D$ cannot hold of $T$ for any $n < b$, so case 2 must hold for all $n < b$. That is, for each $n < b$ there is a $k$ such that for all $\sigma \in T \cap 2^k$, we have $\exists j \leqslant k \, \widehat{\theta}'_\Sigma(n, \sigma \upharpoonright j)$. Since $\mathrm{B}\Sigma^0_1$ holds in $\mathcal{M}$ (by Exercise 4.8), there is an $m$ such that for all $n < b$ and all $\sigma \in T \cap 2^k$, we have $\exists j \leqslant m \, \widehat{\theta}'_\Sigma(n, \sigma \upharpoonright j)$. Thus $\mathcal{M} \cup \{G\} \vDash \exists z \, \forall w < b \, \exists u < z \, \widehat{\theta}'_\Sigma(w, G \upharpoonright u)$.

So we see that in fact

$$\mathcal{M} \cup \{G\} \vDash [\exists z \, \forall w < t \, \exists u < z \, \theta'_\Sigma(G \upharpoonright u)] \leftrightarrow [\forall w < t \, \exists z \, \theta'_\Sigma(G \upharpoonright z)],$$

and hence

$$\mathcal{M} \cup \{G\} \vDash \theta(G) \leftrightarrow \exists z \, \forall w < t \, \exists u < z \, \theta'_\Sigma(G \upharpoonright u),$$

as required.                                                                                    $\square$

We are now ready to show that $\mathcal{M} \cup \{G\} \vDash \mathrm{I}\Sigma^0_1$. We show that $\mathcal{M} \cup \{G\}$ satisfies the least number principle for $\Pi^0_1$ formulas (see the comment before Exercise 4.31). Let $\theta(x, Y)$ be a $\Pi^0_1$ formula with parameters from $\mathcal{M}$ and no free variables other than $x$ and $Y$, such that $\mathcal{M} \cup \{G\} \vDash \exists x \, \theta(x, G)$. We need to show that $I = \{n \in M : \mathcal{M} \cup \{G\} \vDash \theta(n, G)\}$ has a least element. Let $b \in M$ be such that $\mathcal{M} \cup \{G\} \vDash \theta(b, G)$. It is enough to show that $\{n \leqslant_M b : \mathcal{M} \cup \{G\} \vDash \theta(n, G)\}$ is $\mathcal{M}$-finite, as then this set has a least element, which is also the least element of $I$.

Let $\psi(x, \tau)$ be as in Lemma 7.13 for $\varphi \equiv \neg\theta$, and let $D$ be as in Lemma 7.12. Since $D \in \mathcal{D}$, there is a $T \in D$ such that $G$ is a path on $T$. By Exercise 4.7, $\mathcal{M}$ satisfies bounded $\Sigma^0_1$-comprehension, so the set $F$ of all $n \leqslant_M b$ that satisfy item 2 in the definition of $D$ is $\mathcal{M}$-finite. Hence so is $[0, b] \setminus F$, which is exactly the set $\{n \leqslant_M b : \mathcal{M} \cup \{G\} \vDash \theta(n, G)\}$.                                            $\square$

It is easy to see from the proof of Proposition 7.10 that if we have two principles $P$ and $Q$, both of which satisfy the hypothesis of that proposition, then $\mathrm{RCA}_0 + P + Q$ is $\Pi^1_1$-conservative over $\mathrm{RCA}_0$ (of course, the same is true for any number of such principles, even countably infinitely many). Thus, combining the proofs of Exercise 7.11 and Theorem 7.8, we see that $\mathrm{WKL}_0 + \mathrm{COH}$ is $\Pi^1_1$-conservative over $\mathrm{RCA}_0$.

Conservativity results over stronger systems can also be quite interesting, as we saw in Theorems 6.83 and 6.91. Both WKL and COH have been shown to be $\Pi^1_1$-conservative over both $B\Sigma^0_2$ and $I\Sigma^0_2$ (Hájek [74] and Chong, Slaman, and Yang [23]), in addition to being conservative over $RCA_0$. We will see another example of a principle with this property in Section 9.3, as well as another principle that is $\Pi^1_1$-conservative over $RCA_0$ and over $I\Sigma^0_2$, but not over $B\Sigma^0_2$. The following question is closely related to Question 6.86.

▶ **Open Question 7.14.** Is (S)$RT^2_2$ $\Pi^1_1$-conservative over $B\Sigma^0_2$?

## 7.3 COH and r-$\Pi^1_2$-conservativity

Theorems 6.81 and 6.82 show that $SRT^2_2$ implies $B\Sigma^0_2$ over $RCA_0$, and hence $RCA_0 + SRT^2_2$ is not arithmetically conservative over $RCA_0$. Indeed, Cholak, Jockusch, and Slaman [20] showed that $RCA_0 + SRT^2_2$ is not even $\Pi^0_4$-conservative over $RCA_0$. Exercise 7.11 shows that the situation for COH is quite different. In this section, we give an extension of the result in that exercise, which also yields a unified proof that COH does not imply most of the principles we have considered.

**Definition 7.15.** Let r-$\Pi^1_2$ (for *restricted* $\Pi^1_2$) be the collection of sentences of the form

$$\forall X \, [\eta(X) \rightarrow \exists Y \, \rho(X, Y)] \tag{7.2}$$

where $\eta$ is arithmetic and $\rho$ is $\Sigma^0_3$.

Note that if $T_1$ is r-$\Pi^1_2$-conservative over $T_2$, then it is $\Pi^1_1$-conservative over $T_2$, since the $\Pi^1_1$ sentence $\forall X \, \eta(X)$ is equivalent to the r-$\Pi^1_2$ sentence $\forall X \, [\neg\eta(X) \rightarrow \exists Y \, [Y \neq Y]]$.

Examples of principles that can be stated as r-$\Pi^1_2$ formulas include WKL, $RT^2_2$, and $SRT^2_2$ (as well as principles such as ADS and CAC that will be discussed below, and many others). If $RCA_0 + P$ is r-$\Pi^1_2$-conservative over $RCA_0$, then $RCA_0 + P$ does not imply any of these principles. We will show that $RCA_0 + COH$ is r-$\Pi^1_2$-conservative over $RCA_0$. Note that COH itself can be expressed in the form (7.2) where $\varphi$ is $\Pi^0_3$, so this result is tight.

We have the following version of Proposition 7.10.

**Proposition 7.16.** *Let* $P \equiv \forall X \, [\theta(X) \rightarrow \exists Y \, \psi(X, Y)]$, *with* $\theta$ *and* $\psi$ *arithmetic. Suppose that for every countable model* $\mathcal{M}$ *of* $RCA_0$ *and every* $X$ *such that* $\mathcal{M} \models \theta(X)$, *there is a* $G$ *such that*

1. $\mathcal{M} \cup \{G\} \vDash I\Sigma_1^0 + \psi(X, G)$ *and*
2. *for every $\Sigma_3^0$ formula $\rho(Z, Y)$ with no free variables other than $Z$ and $Y$, and every $Z \in S_M$, if $\mathcal{M} \vDash \neg\exists Y \rho(Z, Y)$ then $\mathcal{M}[G] \vDash \neg\exists Y \rho(Z, Y)$.*

*Then $\mathrm{RCA}_0 + P$ is $r$-$\Pi_2^1$-conservative over $\mathrm{RCA}_0$.*

*Proof.* It suffices to take an $r$-$\Pi_2^1$ sentence $Q \equiv \forall X \, [\eta(X) \to \exists Y \rho(X, Y)]$ that fails in some countable model $\mathcal{M}$ of $\mathrm{RCA}_0$ and produce an $\omega$-extension $\mathcal{N}$ of $\mathcal{M}$ such that $\mathcal{N} \vDash \mathrm{RCA}_0 + P + \neg Q$. We build $\mathcal{N}$ as in the proof of Proposition 7.10, using the $G$'s given above, which ensures that $\mathcal{N} \vDash \mathrm{RCA}_0 + P$. There is an $X \in S_M$ such that $\mathcal{M} \vDash \eta(X) \wedge \neg\exists Y \rho(X, Y)$. By induction, $\mathcal{M}_n \vDash \neg\exists Y \rho(X, Y)$ for all $n$ (where $\mathcal{M}_n$ is as in the proof of Proposition 7.10). Since $\eta$ and $\rho$ are arithmetic, $\mathcal{N} \vDash \eta(X) \wedge \neg\exists Y \rho(X, Y)$, so $\mathcal{N} \vDash \mathrm{RCA}_0 + P + \neg Q$. $\qquad\square$

The proof that $\mathrm{RCA}_0 + \mathrm{COH}$ is $r$-$\Pi_2^1$-conservative over $\mathrm{RCA}_0$ relies on the following result.

**59** **Exercise 7.17** (Hirschfeldt and Shore [88]). Let $\mathcal{M}$ be a countable model of $\mathrm{RCA}_0$ and let $G$ be sufficiently generic for the notion of Mathias forcing in $\mathcal{M}$, as described in Exercise 7.11. Show that if $\mathcal{M}[G] \vDash \Theta(G)$, where $\Theta$ is a $\Sigma_3^0$ property with parameters from $\mathcal{M}$, then there is a $Z \in S_M$ such that $\mathcal{M} \vDash \Theta(Z)$. [Hint: Write $\Theta(X)$ as $\exists x \forall y \exists z \, \Phi(X, x, y, z)$ and fix $x \in M$ such that $\mathcal{M}[G] \vDash \forall y \exists z \, \Phi(G, x, y, z)$. By Exercise 2.9, there is a condition $(F_0, I_0)$ such that $(F_0, I_0) \Vdash \forall y \exists z \, \Phi(G, x, y, z)$. Proceed by recursion to obtain $(F_0, I_0) \succcurlyeq (F_1, I_1) \succcurlyeq \cdots$ such that $I_n \setminus I_{n+1}$ is $\mathcal{M}$-finite and $(F_{n+1}, I_{n+1}) \Vdash \exists z \, \Phi(G, x, n, z)$. Show that this procedure can be carried out so that $Z = \bigcup_n F_n$ is $\Delta_1^0$-definable over $\mathcal{M}$, and hence is in $S_M$, and that $\mathcal{M} \vDash \forall y \exists z \, \Phi(Z, x, y, z)$.]

It is useful to think of the recursive definition of the conditions in the above exercise as an effective procedure, and thus think of $Z$ as computable relative to $I_0$ and the parameters of $\Phi$. (See the discussion of computability in $\mathrm{RCA}_0$ in Section 4.2.) For a proof of the following theorem (including a proof of the result in the above exercise) done using computability theoretic notation, see [88].

**Theorem 7.18** (Hirschfeldt and Shore [88]). $\mathrm{RCA}_0 + \mathrm{COH}$ *is $r$-$\Pi_2^1$-conservative over $\mathrm{RCA}_0$.*

*Proof.* Let $\mathcal{M}$ be a countable model of $\mathrm{RCA}_0$. Consider the notion of Mathias forcing in $\mathcal{M}$, as described in Exercise 7.11. If $G$ is sufficiently generic for this notion, then $G$ cohesive for all sets in the second order part of $\mathcal{M}$, by the same argument as in the proof of Proposition 6.41. (Of course, "cohesive" here should be understood in the sense of $\mathcal{M}$; i.e., unbounded and such that for every $X \in S_M$, either $G \cap X$ is $\mathcal{M}$-finite or $G \cap (M \setminus X)$ is $\mathcal{M}$-finite.) By the solution to Exercise 7.11, we also have $\mathcal{M} \cup \{G\} \vDash I\Sigma_1^0$. Thus, by Proposition 7.16, it is enough to show that for every $\Sigma_3^0$ formula $\rho(X, Y)$ with no free variables other than $X$ and $Y$, and every $X \in S_M$, if $\mathcal{M} \vDash \neg \exists Y \, \rho(X, Y)$ then $\mathcal{M}[G] \vDash \neg \exists Y \, \rho(X, Y)$.

Fix such a $\rho$ and $X$, and suppose that $\mathcal{M}[G] \vDash \exists Y \, \rho(X, Y)$. We show that $\mathcal{M} \vDash \exists Y \, \rho(X, Y)$. Let $Y$ be an element of the second order part of $\mathcal{M}[G]$ such that $\mathcal{M}[G] \vDash \rho(X, Y)$. Let $\varphi(x, Z)$ be a $\Sigma_1^0$ formula and $\psi(x, Z)$ a $\Pi_1^0$ formula, both with parameters from $\mathcal{M}$, such that both $\varphi(x, G)$ and $\psi(x, G)$ define $Y$ over $\mathcal{M} \cup \{G\}$. Let $\eta(Z, W) \equiv \forall x \, [x \in W \leftrightarrow \varphi(x, Z) \wedge x \in W \leftrightarrow \psi(x, Z)]$. It is not difficult to see that, by replacing occurrences of $t \in Y$ in $\rho$ by either $\varphi(t, G)$ or $\psi(t, G)$ and rearranging quantifiers, we can obtain a $\Sigma_3^0$ formula $\nu(Z)$ with parameters from $\mathcal{M}$ such that

$$\mathcal{M}[G] \vDash \forall Z \, \forall W \, [\eta(Z, W) \rightarrow [\nu(Z) \leftrightarrow \rho(X, W)]].$$

If $Z \in S_M$ and $\mathcal{M} \vDash \forall x \, [\varphi(x, Z) \leftrightarrow \psi(x, Z)]$, then by $\Delta_1^0$-comprehension, there is a $W \in S_M$ such that $\mathcal{M} \vDash \eta(Z, W)$. If in addition $\mathcal{M} \vDash \nu(Z)$, then $\mathcal{M} \vDash \rho(X, W)$, as required. Thus, letting $\Theta(Z) \equiv \forall x \, [\varphi(x, Z) \leftrightarrow \psi(x, Z)] \wedge \nu(Z)$, it is enough to show that there is a $Z \in S_M$ such that $\mathcal{M} \vDash \Theta(Z)$. But $\Theta$ is a $\Sigma_3^0$ property, and $\mathcal{M}[G] \vDash \Theta(G)$, so the existence of such a $Z$ follows from Exercise 7.17. $\square$

Note that if $P$ is a principle that can be stated as an r-$\Pi_2^1$ formula and fails to hold in some $\omega$-model of $\mathrm{RCA}_0$, then the above proof yields an $\omega$-model of $\mathrm{RCA}_0 + \mathrm{COH} + \neg P$. In particular, we have the following consequence, which also follows from Theorems 6.81 and 6.82 and Exercise 7.11.

**Corollary 7.19** (Cholak, Jockusch, and Slaman [20]). $\mathrm{RCA}_0 + \mathrm{COH} \nvdash \mathrm{SRT}_2^2$.

# Chapter 8

# Drawing a Map: Five Diagrams

Let us summarize some of the results of the previous chapters. In all of the diagrams below, no other implications than the ones shown (or implied by transitivity) hold. Readers may find it useful to go through the following diagrams and justify the implications and nonimplications, using results given above. Expanded versions of some of these diagrams can be found in Hirschfeldt and Jockusch [85]. These diagrams were generated by hand, but there is a tool available for automating the drawing of such diagrams, known as "The Reverse Mathematics Zoo". At the time of writing, it is maintained by Damir Dzhafarov at `rmzoo.uconn.edu`.

Our first diagram, Figure 8.1, shows implications over $RCA_0$, and one open question. If we work over $WKL_0$ instead of $RCA_0$, this diagram remains unchanged except for the obvious collapse of $WKL_0$ and $RCA_0$.

Our second diagram, Figure 8.2, shows implications over RCA (i.e., $RCA_0$ plus full induction), or in the sense of $\omega$-models. There are two arrows with questions marks, but they represent the same open question, since $RT_2^2$ is equivalent to $SRT_2^2 + COH$ over $RCA_0$.

Our third diagram, Figure 8.3, shows implications between the first order parts of various principles, and several open questions. These implications are over $RCA_0$, or equivalently, over $\Sigma_1^0$-PA. Here $(P)^1$ denotes the first order consequences of $RCA_0 + P$. As equality in the first diagram implies equality in this one, some principles are suppressed for clarity.

Our fourth diagram, Figure 8.4, shows computability theoretic implications, i.e. under $\leqslant_c$, and one open question. In light of Question 6.3, we fix $k > 1$ in this diagram.

Our final diagram, Figure 8.5, shows uniform computability theoretic implications, i.e. under $\leqslant_u$. For clarity, we restrict ourselves mostly to 2-colorings.

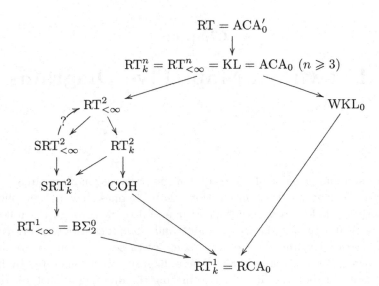

Fig. 8.1   Implications over RCA$_0$

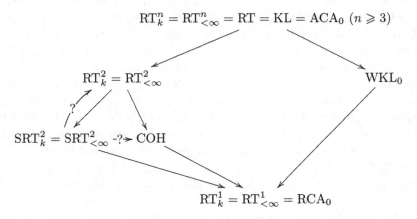

Fig. 8.2   Implications over RCA, or in the sense of $\omega$-models

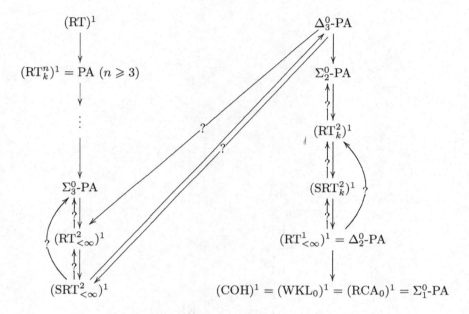

Fig. 8.3   Implications between first order parts

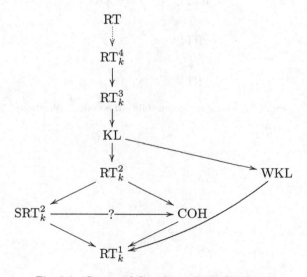

Fig. 8.4   Computability theoretic implications

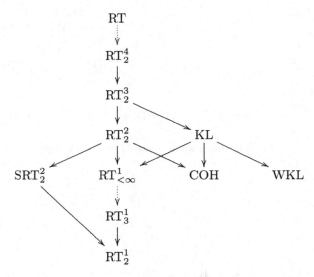

Fig. 8.5   Uniform computability theoretic implications

# Chapter 9

# Exploring Our Surroundings: The World below $\text{RT}_2^2$

Unlike the case of $\text{WKL}_0$, there do not seem to be many principles equivalent to $\text{RT}_2^2$ over $\text{RCA}_0$. There is, however, a whole world of principles, particularly combinatorial ones, *below* $\text{RT}_2^2$ (i.e., provable in $\text{RCA}_0 + \text{RT}_2^2$) in the reverse mathematical universe. We have already discussed $\text{SRT}_2^2$ and COH. In this chapter, we consider a few more, to illustrate some of the workings of this world below $\text{RT}_2^2$. We begin by focusing on a principle called ADS, as an extended example. We then briefly mention several other combinatorial principles provable from $\text{RT}_2^2$. Finally, we discuss some basic theorems that come from what would seem like quite a different area of mathematics, namely model theory, but nevertheless also inhabit this part of the reverse mathematical universe. For the most part, the level of detail in this chapter will be lower than in our long case study of $\text{RT}_2^2$; much more information can be found in the cited papers, among several others. The literature of this area of research continues to expand, so searching for recent papers and preprints is likely to prove rewarding. It is also a useful exercise to fill out the diagrams in the previous chapter with the principles mentioned below, either by hand or using the Reverse Mathematics Zoo mentioned in the previous chapter; several of the papers cited in this chapter also contain such extended diagrams.

## 9.1 Ascending and descending sequences

There are many natural principles that can be seen as special cases of the idea, embodied in Ramsey's Theorem, that large structures must contain large ordered substructures. The following are two of the simplest examples.

Let $\preceq$ be a linear order. An infinite sequence $x_0, x_1, \ldots$ is *ascending* if $x_0 \prec x_1 \prec \cdots$ and *descending* if $x_0 \succ x_1 \succ \cdots$.

**Definition 9.1.** The *Ascending / Descending Sequence Principle* (ADS) is the statement that every infinite linear order has an infinite ascending or descending sequence.

Given a partial order on a set $X$, a *chain* is a set $S \subseteq X$ such that any two elements of $S$ are comparable, and an *antichain* is a set $S \subseteq X$ such that no two elements of $S$ are comparable.

**Definition 9.2.** The *Chain / Antichain Principle* (CAC) is the statement that every infinite partial order has an infinite chain or antichain.

Given a partial order, if we give the pair $\{x, y\}$ the color 0 if $x$ and $y$ are comparable and the color 1 otherwise, then any homogeneous set for this coloring is a chain or antichain of our partial order. Thus $RT_2^2$ implies CAC over $RCA_0$. The first part of the following exercise shows that the same is true of ADS.

**60** **Exercise 9.3** (Hirschfeldt and Shore [88]).
**a.** Show that $RCA_0 + CAC \vdash ADS$.
**b.** Let $\preccurlyeq$ be a linear order on $X \subseteq \mathbb{N}$. A set $S \subseteq X$ *forms an ascending sequence* if for all $x, y \in S$, we have $x < y \rightarrow x \prec y$. A set $S \subseteq X$ *forms a descending sequence* if for all $x, y \in S$, we have $x < y \rightarrow y \prec x$. (Here $<$ is the usual order on natural numbers.) Another natural way to formulate the ADS principle would be to say that for every infinite linear order, there is an infinite set that forms an ascending or descending sequence. Show that this formulation is equivalent to the one in Definition 9.1 over $RCA_0$.

We will discuss ADS in this section, leaving CAC to Section 9.2. We begin by further pinpointing the relationship between ADS and principles discussed in previous chapters.

**Theorem 9.4** (Hirschfeldt and Shore [88]). $RCA_0 + ADS \vdash COH$.

*Proof.* Fix a sequence $R_0, R_1, \ldots$ and define a linear order $\preccurlyeq$ by $x \preccurlyeq y$ iff $(R_i(x))_{i \leqslant x} \leqslant_{\text{lex}} (R_i(y))_{i \leqslant y}$, where $\leqslant_{\text{lex}}$ is the lexicographic order on $2^{<\mathbb{N}}$. Assume there is an infinite set $S$ that forms an infinite ascending sequence for $\preccurlyeq$, in the sense of Exercise 9.3, the descending case being symmetric. Fix $j \in \mathbb{N}$. By bounded $\Sigma_1^0$-comprehension (see Exercise 4.7), we can form the set of all $\sigma$ of length $j+1$ such that $\sigma \leqslant_{\text{lex}} (R_i(x))_{i \leqslant x}$ for some $x \in S$. This set is finite, so it has a lexicographically largest element $\sigma$. Let $x \in S$ be such that $\sigma \leqslant_{\text{lex}} (R_i(x))_{i \leqslant x}$. Let $y \in S$ be larger than $x$ in natural number order. Then $y \succcurlyeq x$, so $\sigma \leqslant_{\text{lex}} (R_i(y))_{i \leqslant y}$, and hence $\sigma \leqslant_{\text{lex}} (R_i(y))_{i \leqslant j}$. By our choice of $\sigma$, we cannot have $\sigma <_{\text{lex}} (R_i(y))_{i \leqslant j}$, so in fact $\sigma = (R_i(y))_{i \leqslant j}$. In particular, $R_j(y)$ has the same value for all $y \in S$ larger than $x$ in natural number order. Thus $S$ is cohesive for $R_0, R_1, \dots$. $\qquad\square$

On the other hand, ADS can be stated as an r-$\Pi_2^1$ sentence, and Theorem 9.4 shows that $RCA_0 \nvdash ADS$ (indeed, it shows that even $WKL_0 \nvdash ADS$), so by Theorem 7.18, $RCA_0 + COH \nvdash ADS$.

The next exercise uses the following theorem to give an alternate proof that $WKL_0 \nvdash ADS$.

**Theorem 9.5** (Herrmann [82]). *There is an infinite computable partial order with no $\Delta_2^0$ infinite chain or antichain.*

$\boxed{61}$ **Exercise 9.6** (Hirschfeldt and Shore [88]). Show that for any computable partial order $\leqslant_P$, there is a computable linear order $\leqslant_L$ extending $\leqslant_P$. (That is, the domain of $\leqslant_L$ is the same as that of $\leqslant_P$, and if $x \leqslant_P y$ then $x \leqslant_L y$.) Then show that, for the partial order $\leqslant_P$ in Herrmann's Theorem, the corresponding linear order $\leqslant_L$ has no low infinite ascending or descending sequence.

It is natural to ask whether ADS implies (S)$RT_2^2$ or $WKL_0$. To discuss the answers to these questions, we take a brief detour of independent interest.

It might seem at this point that the natural consequences of $RT_2^2$ and of WKL form disjoint classes. However, this is not the case, as we now see. Recall from Section 4.2 that DNR is the statement that for each set $A$ there is a function that is diagonally noncomputable relative to $A$ (i.e., a function $f$ such that $f(e) \neq \Phi_e^A(e)$ for all $e$).

$\boxed{62}$ **Exercise 9.7** (Giusto and Simpson [71]). Show that $WKL_0 \vdash DNR$.

Giusto and Simpson [71] showed that in fact $WWKL_0 \vdash DNR$; Ambos-Spies, Kjos-Hanssen, Lempp, and Slaman [2] showed that $RCA_0 + DNR \nvdash WWKL_0$.

As mentioned in Section 3.1, every $0, 1$-valued diagonally noncomputable function has PA degree. General diagonally noncomputable functions, on the other hand, are computability theoretically weaker. The reverse mathematical consequence of this fact is that DNR is weaker than $WKL_0$. For example, we have the following result.

**Theorem 9.8** (Hirschfeldt, Jockusch, Kjos-Hanssen, Lempp, and Slaman [86]). $\mathrm{RCA}_0 + \mathrm{SRT}_2^2 \vdash \mathrm{DNR}$.

Thus we see that DNR follows from both WKL and $\mathrm{RT}_2^2$. The connection with ADS comes from the following result.

**Theorem 9.9** (Hirschfeldt and Shore [88]). $\mathrm{RCA}_0 + \mathrm{ADS} \nvdash \mathrm{DNR}$.

**Corollary 9.10** (Hirschfeldt and Shore [88]). ADS *implies neither* $\mathrm{SRT}_2^2$ *nor* $\mathrm{WKL}_0$ *over* $\mathrm{RCA}_0$.

We will say a word about the proof of Theorem 9.9 later in this section.

We now move to some computability theoretic results, before returning to reverse mathematics. It follows from Theorems 6.34 and 9.4 that ADS is not effectively true, but the following direct proof is worth knowing, as it is a particularly clean way to introduce the priority method, which was created by Friedberg [57] and Muchnik [150], and has proved invaluable in virtually all branches of computability theory and its applications. The version of the priority method in this proof (as in the original Friedberg-Muchnik argument) is known as finite injury, as explained below. For more on priority arguments, including infinite injury arguments, see for instance Soare [196].

**Theorem 9.11** (Tennenbaum, see Rosenstein [172]; Denisov, see Goncharov and Nurtazin [72]). *There is a computable infinite linear order with no computable infinite ascending or descending sequence.*

*Proof.* We build a computable linear order $(\mathbb{N}, \prec)$. We will ensure that no infinite c.e. set can be the range of an ascending or descending sequence. We break our task up into requirements

$$R_e : W_e \text{ infinite } \Rightarrow \exists x, y \in W_e \,[x \text{ has finitely many } \prec\text{-predecessors}$$
$$\land \; y \text{ has finitely many } \prec\text{-successors}].$$

For each $e$, we will have numbers $x_e \prec y_e$, which are initially undefined and may change value during the construction, which we will try to make into witnesses to the satisfaction of $R_e$ by placing all sufficiently large numbers into the $\prec$-interval $(x_e, y_e)$.

For those unfamiliar with priority arguments, the following description of the construction may be helpful: We proceed in stages, at each stage $s$ determining the position of $s$ in our order. For each $e$, we can wait until some pair of numbers $x_e \prec y_e$ enters $W_e$ at some stage $s$, which must happen if $W_e$ is infinite, and then try to place every $t \geqslant s$ into the $\prec$-interval $(x_e, y_e)$. However, this action could create a conflict if we have, say, $y_i \prec x_e$ for some $i$. An attempt to resolve this situation is to insist that, when we define $x_e$ and $y_e$, we do so only if we have $x_i \preccurlyeq x_e \prec y_e \preccurlyeq y_i$ for every $i$ such that $x_i$ and $y_i$ have been previously defined. However, we might now not be able to define $x_e$ and $y_e$ even if $W_e$ is infinite. For instance, suppose we define $x_1$ and $y_1$, and $x_1$ then enters $W_0$. Then we define $x_2$ and $y_2$, and $x_2$ then enters $W_0$. Next, we define $x_3$ and $y_3$, and $x_2$ then enters $W_0$. If this pattern keeps occurring, then we can never define $x_0$ and $y_0$ (and indeed $W_0$ forms an infinite ascending sequence). The solution to this problem is to assign *priorities* to our requirements. Requirement $R_0$ has the strongest priority, so we define $x_0$ and $y_0$ as soon as we can, regardless of their positions in relation to previously defined $x_i$ and $y_i$. To avoid being back in the original situation where independent definitions of the $x_e$ and $y_e$ caused problems, we undefine all $x_i$ and $y_i$ for $i > 0$. Requirement $R_1$ has priority over all other requirements *except* $R_0$. Once $x_0$ and $y_0$ are defined, we will not define $x_1$ and $y_1$ unless we can ensure that $x_0 \prec x_1 \prec y_1 \prec x_1$. In general, when we define $x_e$ and $y_e$, we ensure that $x_i \prec x_e \prec y_e \prec y_i$ for all $i < e$ such that $x_i$ and $y_i$ are defined, and undefine $x_j$ and $y_j$ for all $j > e$. In this case, we say that $R_j$ is *injured* for each such $j$. It is easy to see that $R_0$ is never injured, $R_1$ is injured at most once, $R_2$ is injured at most three times, and in general $R_e$ is injured only finitely often. Thus this construction is a *finite injury priority* construction, which ensures that each requirement is eventually met.

We now proceed with our construction. At stage $s$, we say that $R_e$ *requires attention* if $x_e$ and $y_e$ are currently undefined and there are numbers $x, y \in W_e[s]$ such that $x \prec y$ and for every $i < e$ for which $x_i$ and $y_i$ are defined, $x_i \prec x \prec y \prec y_i$. (We adopt the usual convention that if $x, y \in W_e[s]$, then $x, y < s$, so the positions of $x$ and $y$ in our order relative to each other and other numbers $< s$ have already been determined.) If there is some $e < s$ that requires attention, then for the least such $e$, we let $x$ and $y$ be as above and define $x_e = x$ and $y_e = y$. We also undefine $x_i$ and $y_i$ for $i > e$. In any case, we now place $s$ into our order so that, for all $i$ for which $x_i$ and $y_i$ are currently defined, $x_i \prec s \prec y_i$.

Clearly the above construction can be carried out so as to build a computable linear order. We now show by induction that every $R_e$ eventually stops requiring attention and is met. Assume that this statement is true for all $i < e$, and let $s$ be a stage by which all $R_i$ with $i < e$ have stopped requiring attention. Then $R_e$ can require attention at most once at a stage

$\geqslant s$. If $W_e$ is finite then $R_e$ is vacuously true, so suppose $W_e$ is infinite. For all $i < e$, either $x_i$ and $y_i$ are permanently undefined from stage $s$ on, or the values of $x_i$ and $y_i$ at stage $s$ are their final values. In the latter case, every $t \geqslant s$ is placed in the $\prec$-interval $[x_i, y_i]$, so if $x_e$ and $y_e$ are not defined at stage $s$, they must become defined at some stage $\geqslant s$, since we are assuming $W_e$ is infinite. Let $t \geqslant s$ be a stage at which $x_e$ and $y_e$ are defined. Then every $u \geqslant t$ is placed in the $\prec$-interval $[x_e, y_e]$, so $R_e$ is met.                                                                                           □

We denote the order type of an infinite ascending sequence by $\omega$, and the inverse of that type (i.e., the order type of an infinite descending sequence) by $\omega^*$. As usual, for order types $\mu$ and $\nu$, by $\mu + \nu$ we mean the order type of an order of type $\mu$ followed by one of type $\nu$, and by $\mu\nu$ we mean the order type of an order obtained by replacing each element of an order of type $\nu$ by one of type $\mu$.

$\boxed{63}$ **Exercise 9.12** (Tennenbaum, see [172]; Denisov, see [72]). Modify the proof of Theorem 9.11 to ensure that the linear order being constructed has order type $\omega + \omega^*$ (or verify that the proof already does so).

$\boxed{64}$ **Exercise 9.13** (Lerman [124]). Show that there is a computable infinite linear order with no computable suborder of type $\omega$, $\omega^*$, or $\omega + \omega^*$. [Hint: This is another finite injury priority construction. The requirements are now

$$R_e : W_e \text{ infinite} \Rightarrow \exists x \in W_e \, [x \text{ has infinitely many } \prec\text{-predecessors}$$

$$\text{and infinitely many } \prec\text{-successors in } W_e].$$

The basic strategy for $R_e$ is to wait until some $x$ enters $W_e$, then start putting all numbers to the left of $x$, until some $y_0 \prec x$ enters $W_e$. We then start putting all numbers to the right of $x$ until some $z_0 \succ x$ enters $W_e$, then switch back to putting all numbers to the left of $x$, until some $y_1 \prec x$ with $y_1 \neq y_0$ enters $W_e$, and so on. Call $x$ a *pivot point* for $R_e$. A weaker priority $R_i$ cannot know whether $W_e$ is infinite, and thus whether the construction might get stuck permanently to the left or to the right of $x$, so it needs to try to find two pivot points, one to the left of $x$, and one to the right of $x$.]

$\boxed{65}$ **Exercise 9.14** (Rosenstein [172]). Show that if a computable infinite linear order has no computable suborder of type $\omega$, $\omega^*$, or $\omega + \omega^*$, then it has order type $\omega + \zeta\eta + \omega^*$. Here $\zeta$ is the order type of the integers and

$\eta$ is the order type of the rationals (so $\zeta\eta$ is densely many copies of the integers).

The special role that orders of type $\omega$, $\omega^*$, or $\omega + \omega^*$ play in the computability theoretic analysis of linear orders can be explained by thinking of a linear order $\preccurlyeq$ on $\mathbb{N}$ as a 2-coloring in the obvious way (i.e., where for $m < n$, we give $\{m, n\}$ color 0 if $m \prec n$ and color 1 otherwise). Then $\preccurlyeq$ is stable iff every element has either finitely many predecessors or finitely many successors, which means that $\preccurlyeq$ must have order type $\omega$, $\omega^*$, $\omega + \omega^*$, $\omega + n$, or $n + \omega^*$ (where $n$ is the finite linear order with $n$ elements). Since we can always obtain the initial segment of type $\omega$ from an order of type $\omega + n$ computably (and similarly for $n + \omega^*$), we can restrict attention to $\omega$, $\omega^*$, and $\omega + \omega^*$. The following definition captures the notion of a linear order having one of these three order types in a way that is easy to work with in the reverse mathematical setting.

**Definition 9.15.** A linear order is *discrete* if every element has an immediate predecessor, except for the first element of the order if there is one, and every element has an immediate successor, except for the last element of the order if there is one. A linear order is *stable* if it is discrete and has more than one element, and every element has either finitely many predecessors or finitely many successors. (Note that a stable order must be infinite.)

The decomposition of RT$_2^2$ into SRT$_2^2$ and COH, as well as the computability theoretic results above, suggest the idea of decomposing ADS into stable and cohesive parts.

**Definition 9.16.** *Stable ADS* (SADS) is the statement that every stable linear order has an infinite ascending or descending sequence.

*Cohesive ADS* (CADS) is the statement that every infinite linear order has a stable suborder.

An important feature of the notion of stability for colorings is that computable stable colorings can be identified with $\Delta_2^0$ sets, as we have seen. In a computable stable linear order, the set $S$ of elements with only finitely many predecessors (the $\omega$ part of a linear order of type $\omega + \omega^*$) is also a $\Delta_2^0$ set, and the set of elements with only finitely many successors is its complement. The difference, of course, is that in this case there is an order that places some elements "deeper inside" $S$ or $\overline{S}$ than others. Indeed, one may think of ADS/SADS/CADS as the linearly ordered versions of RT$_2^2$/SRT$_2^2$/COH, and hence of the study of the differences between these sets of principles as a case study in the computability theoretic and reverse mathematical effects of imposing a linear order on a structure.

It is easy to see that $\mathrm{RCA}_0 \vdash \mathrm{ADS} \leftrightarrow \mathrm{SADS} + \mathrm{CADS}$. Exercises 9.12 and 9.13 show that $\mathrm{RCA}_0$ proves neither SADS nor CADS. As might be expected, we also have the following implications.

**66** **Exercise 9.17** (Hirschfeldt and Shore [88]). Show that $\mathrm{RCA}_0 + \mathrm{SRT}_2^2 \vdash \mathrm{SADS}$ and $\mathrm{RCA}_0 + \mathrm{COH} \vdash \mathrm{CADS}$.

CADS may not look much like COH at first glance, but the solution to Exercise 6.31 can easily be adapted to show that COH implies *Cohesive Ramsey's Theorem for Pairs* ($\mathrm{CRT}_2^2$), which states that every 2-coloring of pairs of natural numbers is stable on some infinite subset of $\mathbb{N}$. It is open whether $\mathrm{CRT}_2^2$ implies COH over $\mathrm{RCA}_0$, but Hirschfeldt and Shore [88] showed that $\mathrm{CRT}_2^2$ does imply COH over $\mathrm{RCA}_0 + \mathrm{B}\Sigma_2^0$. CADS is the analog to $\mathrm{CRT}_2^2$ for ADS. Hirschfeldt and Shore [88] showed that CADS, too, implies COH over $\mathrm{RCA}_0 + \mathrm{B}\Sigma_2^0$, but the following question is also still open.

▶ **Open Question 9.18.** Does $\mathrm{RCA}_0 + \mathrm{CADS} \vdash \mathrm{COH}$?

It is also open whether $\mathrm{RCA}_0 + \mathrm{CADS} \vdash \mathrm{CRT}_2^2$.

Just as in the case of ADS, we have $\mathrm{RCA}_0 + \mathrm{COH} \nvdash \mathrm{SADS}$ by r-$\Pi_2^1$-conservativity, and hence $\mathrm{RCA}_0 + \mathrm{CADS} \nvdash \mathrm{SADS}$. In the other direction, we have the following result.

**Theorem 9.19** (Hirschfeldt and Shore [88]). *Every computable linear order of type $\omega + \omega^*$ has a low suborder of type $\omega$ or $\omega^*$.*

**67** **Exercise 9.20** (Hirschfeldt and Shore [88]). Prove Theorem 9.19 as follows. Let $\preceq$ be a linear order on the natural numbers, of type $\omega + \omega^*$. Assume that $\preceq$ has no computable infinite descending sequence. Let $A$ consist of all numbers in the $\omega$ part of this order. Consider the notion of forcing whose conditions are finite sequences of elements of $A$ that are increasing in both natural order and $\preceq$-order, where extension is as usual for finite sequences. For each $i$, let $C_i$ be the set of $\rho \in \mathbb{N}^{<\mathbb{N}}$ such that $\rho$ is strictly ascending in both natural number order and $\preceq$, and $\Phi_i^\rho(i)\downarrow$ (meaning that $\Phi_i^{\mathrm{rng}\,\rho}(i)\downarrow$ with use greater than the last element of $\rho$). Define an $\emptyset'$-computable sequence of conditions $\sigma_0, \sigma_1, \ldots$ so that for every $i$ either $\Phi_i^{\sigma_i}(i)\downarrow$ or there is no extension $\rho \in C_i$ of $\sigma_i$. Obtain a low infinite ascending sequence from this sequence.

Relativizing this result and applying Proposition 4.27 in the usual way, we get a model of RCA$_0$ + SADS consisting entirely of low sets. Indeed, applying Exercise 4.28, we get a model of WKL$_0$ + SADS consisting entirely of low sets. On the other hand, the computable linear order $\leqslant_L$ constructed in Exercise 9.6 cannot have a low stable suborder, as then an application of the relativized version of Theorem 9.19 would yield a low infinite ascending or descending sequence in $\leqslant_L$. Thus we have the following consequences.

**Corollary 9.21** (Hirschfeldt and Shore [88]). WKL$_0$ + SADS $\nvdash$ CADS *and* WKL$_0$ + SADS $\nvdash$ SRT$_2^2$.

Thus SADS and CADS form a proper decomposition of ADS (even at the level of $\omega$-models, where the corresponding fact is open for SRT$_2^2$ and COH as a decomposition of RT$_2^2$).

Since COH does not imply WKL$_0$, neither does CADS, so CADS and WKL$_0$ are independent. Hirschfeldt and Shore [88] showed the same is true of SADS. Their proof that SADS does not imply WKL$_0$ (which of course also follows from Theorem 6.71) uses the following result.

**Theorem 9.22** (Hirschfeldt and Shore [88]). *Every computable linear order of type $\omega + \omega^*$ has an infinite ascending or descending sequence that does not compute a diagonally noncomputable function.*

$\boxed{68}$ **Exercise 9.23** (Hirschfeldt and Shore [88]). Prove Theorem 9.22 as follows. Let $\preccurlyeq$ be a linear order on the natural numbers, of type $\omega + \omega^*$, and consider the notion of forcing defined in Exercise 9.20. Let $G$ be sufficiently generic for this notion of forcing, and note that $G$ is an ascending sequence in $\preccurlyeq$. Suppose that $\Phi_e^G$ is diagonally noncomputable. Fix a condition $\sigma$ such that $\sigma \vDash \Phi_e^G$ total $\wedge\; \forall i\,[\Phi_i(i){\downarrow} \;\rightarrow\; \Phi_e^G(i) \neq \Phi_i(i)]$. Use $\sigma$ to show that $\preccurlyeq$ has a computable descending sequence, as otherwise there would be a computable diagonally noncomputable function.

Relativizing this result and applying Proposition 4.27, we get a model of RCA$_0$ + SADS containing no diagonally noncomputable functions, thus showing that RCA$_0$ + SADS $\nvdash$ DNR, and hence RCA$_0$ + SADS $\nvdash$ WKL$_0$. The forcing construction in Exercise 9.23 can be combined with that in the proof of Theorem 7.18 to show that RCA$_0$ + SADS + COH $\nvdash$ DNR, whence RCA$_0$ + ADS $\nvdash$ DNR. Since both SRT$_2^2$ and WKL$_0$ imply DNR, this argument yields a proof of Theorem 9.10, demonstrating that, as in the case of RT$_2^2$, the decomposition of ADS into stable and cohesive parts is a fruitful technique.

The first order consequences of ADS have also been studied. Hirschfeldt and Shore [88] showed that $\mathrm{RCA}_0 + \mathrm{ADS} \vdash \mathrm{B}\Sigma^0_2$. This result was improved by Chong, Lempp, and Yang [22], who showed that $\mathrm{RCA}_0 + \mathrm{SADS} \vdash \mathrm{B}\Sigma^0_2$. Their method involves an analysis of the first order part of models of $\mathrm{RCA}_0$ in which $\mathrm{B}\Sigma^0_2$ fails; in particular, they introduce the notion of a *bi-tame cut*, and show that the existence of such a cut is equivalent over $\mathrm{RCA}_0$ to the failure of $\mathrm{B}\Sigma^0_2$. Conversely, Chong, Slaman, and Yang [23] showed that $\mathrm{RCA}_0 + \mathrm{ADS}$ is $\Pi^1_1$-conservative over $\mathrm{RCA}_0 + \mathrm{B}\Sigma^0_2$, thus pinpointing the first order part of ADS.

We finish this section with an application of Theorem 9.19, illustrating the fact that the relationship between "pure computability theory" and computable mathematics is a two-way street. It is not difficult to see that the $\omega$ part of any computable linear order of type $\omega + \omega^*$ is $\Delta^0_2$. Conversely, Harizanov [78] showed that every $\Delta^0_2$ degree contains the $\omega$ part of some computable linear order of type $\omega + \omega^*$. Jockusch (see [88]) noted that we can therefore argue as follows. Fix a $\Delta^0_2$ degree $\mathbf{a}$, and let $\preccurlyeq$ be a linear order of type $\omega + \omega^*$ whose $\omega$ part $A$ has degree $\mathbf{a}$. Clearly, the $\omega^*$ part $\overline{A}$ of $\preccurlyeq$ also has degree $\mathbf{a}$. It is easy to see that from an infinite ascending sequence in $\preccurlyeq$ we can computably enumerate $A$, while from an infinite descending sequence in $\preccurlyeq$ we can computably enumerate $\overline{A}$. Applying Theorem 9.19, we conclude that every $\Delta^0_2$ degree contains a set that is c.e. in some low degree.

## 9.2  Other combinatorial principles provable from $\mathrm{RT}^2_2$

In this section, we give a small sampling of principles living in the world below $\mathrm{RT}^2_2$.

### 9.2.1  Chains and antichains

We briefly mentioned the Chain / Antichain Principle CAC in the previous section. Hirschfeldt and Shore [88] studied CAC along similar lines to the analysis of ADS above. As in the case of ADS, there are stable and cohesive versions of CAC. The ones given in [88] are as follows. An element $x$ of a partial order $\preccurlyeq$ is said to be *small* if $x \preccurlyeq y$ for all but finitely many $y$, *large* if $y \preccurlyeq x$ for all but finitely many $y$, and *isolated* if $x$ is $\preccurlyeq$-incomparable with all but finitely many $y$. A partial order is *stable* if either each element is small or isolated, or each element is large or isolated. (Notice that in a computable stable partial order, the set of isolated elements is $\Delta_2^0$; cf. the comment following Definition 9.16.)

**Definition 9.24.** *Stable CAC* (SCAC) is the statement that every infinite stable partial order has an infinite chain or antichain.

*Cohesive CAC* (CCAC) is the statement that every infinite partial order has an infinite stable suborder.

CCAC turns out to be a familiar principle in disguise.

⎡69⎤ **Exercise 9.25** (Hirschfeldt and Shore [88]). Show that CCAC and ADS are equivalent over RCA$_0$.

The above notion of stability is not necessarily the most intuitively natural one. (See [88] for more on this choice of definition.) Jockusch, Kastermans, Lempp, Lerman, and Solomon [101] defined a partial order to be *weakly stable* if each element is small, large, or isolated. As the following theorems show, these two notions of stability are computability theoretically different but yield equivalent versions of stable CAC.

**Theorem 9.26** (Jockusch, Lerman, and Solomon, see [101]). *Every infinite computable stable partial order has either an infinite computable chain or an infinite* $\Pi_1^0$ *antichain. On the other hand, there is an infinite computable weakly stable partial order with no infinite* $\Pi_1^0$ *chains or antichains.*

**Theorem 9.27** (Jockusch, Kastermans, and Lempp, see [101]). SCAC *is equivalent over* RCA$_0$ *to the statement that every infinite weakly stable partial order has an infinite chain or antichain.*

Jockusch, Kastermans, and Lempp (see [101]) also showed that, in contrast to the first part of Theorem 9.26, there is an infinite computable stable partial order with no infinite $\Pi_1^0$ chain and no infinite computable antichain. Other related results, as well as useful comments on the notion of stability, can be found in [101].

Wright [personal communication] has noted that the weak stability version of CCAC, i.e., the statement that every infinite partial order has an infinite weakly stable suborder, is equivalent over RCA$_0$ to the principle CRT$_2^2$ introduced in the previous section.

▶ **Open Question 9.28.** What is the reverse mathematical strength of the statement that every infinite weakly stable partial order has an infinite stable suborder?

By Exercise 9.25, the above statement follows from ADS, and it is easy to see that it implies SADS. Wright [personal communication] has shown that, like SADS, it has low solutions, and hence cannot imply ADS.

It is straightforward to adapt the solutions to Exercises 9.20 and 9.23 to obtain $\omega$-models of SCAC consisting entirely of low sets and of sets that do not compute diagonally noncomputable functions, respectively. In the latter case, we can make the model satisfy CAC as in the previous section, so Corollaries 9.10 and 9.21 hold for CAC in place of ADS.

Recently, Lerman, Solomon, and Towsner [125] were able to separate CAC from ADS and SCAC from SADS. Indeed, they obtained the following result.

**Theorem 9.29** (Lerman, Solomon, and Towsner [125]). *There is an $\omega$-model of* $\mathrm{RCA}_0 + \mathrm{ADS}$ *that is not a model of* SCAC.

Their proof has two parts: First, a *ground forcing* construction is used to build an instance $I$ of SCAC with no $I$-computable solutions and an additional special property $P$. Then an $\omega$-model is built in steps using an *iteration forcing* construction, which at each step starts with an $X$ that does not compute any solutions to $I$ and has property $P$ (the base case being $X = I$, of course) and defines a solution $G$ to a given instance of ADS so that $X \oplus G$ also does not compute any solutions to $I$ and has property $P$. This property $P$ is defined so as to make certain subsets of the relevant notion of forcing dense. Among these dense sets are the ones needed to ensure that $X \oplus G$ does not compute any solutions to $I$, and that it has property $P$.

### 9.2.2   Tournaments

Another result in the paper by Lerman, Solomon, and Towsner [125] concerns the following principle. A *tournament* is an irreflexive binary relation $R$ on $\mathbb{N}$ such that for all $x \neq y$, either $R(x, y)$ or $R(y, x)$, but not both. We think of $R$ as representing the results of a round-robin tournament, with $R(x, y)$ iff $x$ beat $y$. The *Erdős-Moser principle* (EM) states that for every tournament $R$, there is an infinite set $S$ such that $R$ is transitive on $S$. The following fact has been noted by several people.

$\boxed{70}$ **Exercise 9.30.** Show that $\mathrm{RT}_2^2$ is equivalent to EM+ADS over $\mathrm{RCA}_0$.

While there seemed to be no way to obtain $\mathrm{RT}_2^2$ from EM without ADS, separating EM from $\mathrm{RT}_2^2$ proved difficult. Indeed, the usual classical proofs of EM in the countable case are actually proofs of $\mathrm{RT}_2^2$, and Kach, Lerman, Solomon, and Weber (see [125]) showed that, as in the case of $\mathrm{RT}_2^2$ (Theorem 6.11), there is a computable instance of EM with no $\Sigma_2^0$ solution. This situation is an excellent example of a good reverse mathematical question: known proofs of EM have a particular combinatorial core, and we want to know whether this core (in this case, the appeal to $\mathrm{RT}_2^2$) is actually necessary. This question was recently answered as follows, with methods similar to those in the proof of Theorem 9.29.

**Theorem 9.31** (Lerman, Solomon, and Towsner [125]). *There is an $\omega$-model of* $\mathrm{RCA}_0 + \mathrm{EM}$ *that is not a model of* $\mathrm{SRT}_2^2$.

Exercise 9.30 can easily be adapted to show that $\mathrm{SRT}_2^2$ follows from $\mathrm{EM} + \mathrm{SADS}$ over $\mathrm{RCA}_0$, so the model in this theorem is also not a model of SADS. For more on the reverse mathematics of EM, including its relationship with one of the model theoretic principles we will discuss in Section 9.3, see [125].

### 9.2.3  Free sets and thin sets

Friedman [61] (see also Friedman and Simpson [66]) introduced the reverse mathematical study of the following Ramsey theoretic principles.

**Definition 9.32.** The *Free Set Theorem* (FS) is the statement that for every $n$ and every $f : [\mathbb{N}]^n \to \mathbb{N}$, there is an infinite $A \subseteq \mathbb{N}$ such that for all $s \in [A]^n$, either $f(s) \notin A$ or $f(s) \in s$. Such an $A$ is known as a *free set* for $f$.

The *Thin Set Theorem* (TS) is the statement that for every $n$ and every $f : [\mathbb{N}]^n \to \mathbb{N}$, there is an infinite $A \subseteq \mathbb{N}$ such that $f([A]^n) \neq \mathbb{N}$. Such an $A$ is known as a *thin set* for $f$.

For a fixed $n$, we denote by $\mathrm{FS}(n)$ the statement that FS holds for $n$, and by $\mathrm{TS}(n)$ the statement that TS holds for $n$.

The reverse mathematics of these and related principles has been studied by Cholak, Giusto, Hirst, and Jockusch [19], but there are still many interesting open questions in this line of research, whose solutions may well require major methodological advances like the one that was required to prove Liu's Theorem 6.71.

$\boxed{71}$ **Exercise 9.33** (Friedman [61]; Cholak, Giusto, Hirst, and Jockusch [19]). Show that for every $n \geqslant 1$, the following are provable in $\mathrm{RCA}_0$.

**a.** FS(1).

**b.** FS($n$) $\to$ TS($n$).

**c.** FS($n + 1$) $\to$ FS($n$) and TS($n + 1$) $\to$ TS($n$).

**d.** TS($n$) is equivalent to the statement that for each $f : [\mathbb{N}]^n \to \mathbb{N}$, there is an infinite $A \subseteq \mathbb{N}$ such that $f([A]^n)$ is coinfinite.

▶ **Open Question 9.34.** Does TS($n + 1$) imply FS($n$) over RCA$_0$?

Friedman [61] gave the following argument. Given $f : [\mathbb{N}]^n \to \mathbb{N}$, let $g : [\mathbb{N}]^{n+1} \to n + 2$ be defined as follows. For $a_1 < \cdots < a_{n+1}$, let $g(a_1, \ldots, a_{n+1})$ be the least $i$ such that $f(a_1, \ldots, a_{i-1}, a_{i+1}, \ldots, a_{n+1}) = a_i$ if such an $i$ exists, and $g(a_1, \ldots, a_{n+1}) = 0$ otherwise. Let $H$ be an infinite homogeneous set for $g$. Then $H$ cannot be homogeneous to any $i > 0$, since otherwise we could take elements $a_0 < \cdots < a_{i-1} < b < c < a_{i+1} < \cdots < a_{n+1}$ of $H$ and conclude both that $f(a_1, \ldots, a_{i-1}, a_{i+1}, \ldots, a_{n+1}) = b$ and that $f(a_1, \ldots, a_{i-1}, a_{i+1}, \ldots, a_{n+1}) = c$. Thus $H$ is homogeneous to 0, and it is easy to see that this fact implies that $H$ is a free set for $f$. So FS($n$) follows from RT$^{n+1}_{n+2}$, and hence ACA$_0$ $\vdash$ FS($n$) for each $n \in \omega$. Similarly, FS follows from RT, so ACA$'_0$ $\vdash$ FS (see Theorem 6.27).

For the $n = 2$ case, a more complicated argument yields the following result.

**Theorem 9.35** (Cholak, Giusto, Hirst, and Jockusch [19]). RCA$_0$ + RT$^2_2$ $\vdash$ FS(2).

In some computability theoretic ways, FS and TS behave much like RT.

$\boxed{72}$ **Exercise 9.36** (Cholak, Giusto, Hirst, and Jockusch [19]). Adapt the proof of Theorem 6.11 and the solution to Exercise 6.15 to show that for each $n \geqslant 2$ there is a computable function $f : [\mathbb{N}]^n \to \mathbb{N}$ with no infinite $\Sigma^0_n$ thin set.

Thus, for any $n \geqslant 2$, TS($n$) does not hold in any $\omega$-model of WKL$_0$ consisting entirely of low sets, and hence WKL$_0 \nvdash$ TS($n$).

**Theorem 9.37** (Cholak, Giusto, Hirst, and Jockusch [19]). *For every $n \geqslant 1$, every computable $f : [\mathbb{N}]^n \to \mathbb{N}$ has an infinite $\Pi^0_n$ free set.*

Recently, Wang [207] made an important advance in the study of these principles by proving the following theorem.

**Theorem 9.38** (Wang [207]). RCA$_0$+FS $\nvdash$ ACA$_0$, *and hence* RCA$_0$+TS $\nvdash$ ACA$_0$.

The following are some of the remaining questions in this line of research.

▶ **Open Question 9.39.** 1. Let $n \geqslant 3$. Does FS($n$) or TS($n$) or FS or TS imply WKL$_0$ over RCA$_0$?

2. Does FS(2) imply RT$_2^2$ (or SRT$_2^2$, COH, ADS, or CAC) over RCA$_0$? What about over WKL$_0$?

3. Does FS(2) imply B$\Sigma_2^0$ over RCA$_0$?

Rice [169] recently showed that TS(2) implies DNR over RCA$_0$.

## 9.2.4 The finite intersection principle and $\Pi_1^0$-genericity

Combinatorial principles of interest to computability theory and reverse mathematics arise in many areas of mathematics. Choice principles in reverse mathematics are discussed in Section VII.6 of Simpson [191], and versions of the axiom of choice related to well-orders have been studied by Friedman and Hirst [63] and Hirst [92]. Dzhafarov and Mummert [51] studied combinatorial equivalents of the axiom of choice, finding several examples that are either equivalent to ACA$_0$ or strictly stronger. In [52], they studied a principle of this sort that lives in the world below RT$_2^2$ (at least when working with a slight amount of additional induction), known as the finite intersection principle (FIP). (They also studied other families of related principles, but we will focus on FIP here.)

A *family of sets* is a sequence $A_0, A_1, \ldots$ of sets of natural numbers. A family of sets $B_0, B_1, \ldots$ is a *subfamily* of $A_0, A_1, \ldots$ if for each $i$ there is a $j$ such that $B_i = A_j$. A family of sets is *nontrivial* if at least one of its members is nonempty. A family of sets has the *finite intersection property* if any intersection of finitely many of its members is nonempty. Given a family of sets $\mathcal{A}$, we say that a subfamily $\mathcal{B}$ of $\mathcal{A}$ is a *maximal subfamily with the finite intersection property* if $\mathcal{B}$ has the finite intersection property and, for any subfamily $\mathcal{C}$ of $\mathcal{A}$ with the finite intersection property, if $\mathcal{B}$ is a subfamily of $\mathcal{C}$ then $\mathcal{C}$ is also a subfamily of $\mathcal{B}$ (i.e., $\mathcal{B}$ and $\mathcal{C}$ have the same members).

**Definition 9.40.** The *Finite Intersection Principle* (FIP) is the statement that every nontrivial family of sets has a maximal subfamily with the finite intersection property.

Note that for a subfamily $\mathcal{B}$ of $A_0, A_1, \ldots$, we cannot always effectively tell which $A_i$ are in $\mathcal{B}$. Strengthening the definition of subfamily to ensure that we can yields a less reverse mathematically interesting principle than

FIP.

73  **Exercise 9.41** (Dzhafarov and Mummert [52]). Show that $\mathrm{ACA}_0$ is equivalent over $\mathrm{RCA}_0$ to the statement that every nontrivial family of sets $A_0, A_1, \ldots$ has a maximal subfamily $\mathcal{B}$ with the finite intersection property for which there is a set $I$ with $i \in I$ iff $A_i \in \mathcal{B}$.

FIP itself, on the other hand, is quite far from $\mathrm{ACA}_0$, as the following theorem shows.

**Theorem 9.42** (Dzhafarov and Mummert [52]). *Let $\mathcal{A}$ be a computable nontrivial family of sets and let $X$ be a noncomputable c.e. set. Then $X$ computes a maximal subfamily of $\mathcal{A}$ with the finite intersection property.*

This theorem shows that even a small amount of noncomputable information can be enough to compute solutions to instances of FIP, and hence raises the obvious question of whether FIP might be effectively true. The answer to this question is negative, and involves an important computability theoretic notion we have already encountered. Recall that a set $X$ is of hyperimmune-free degree if for every $f \leqslant_\mathrm{T} X$, there is a computable $g$ such that $f(n) < g(n)$ for all $n$, and that otherwise $X$ is of hyperimmune degree.

**Theorem 9.43** (Dzhafarov and Mummert [52]). *There is a computable nontrivial family of sets such that any maximal subfamily with the finite intersection property has hyperimmune degree.*

In particular, FIP does not hold in the minimal $\omega$-model of $\mathrm{RCA}_0$, so $\mathrm{RCA}_0 \nvdash \mathrm{FIP}$. On the other hand, Theorem 9.42 allows us to build very "small" $\omega$-models of FIP in a way already mentioned in Chapter 5. Let $X$ be any noncomputable c.e. set. By the Sacks Density Theorem [174], there are c.e. sets $\emptyset = X_0 <_\mathrm{T} X_1 <_\mathrm{T} X_2 <_\mathrm{T} \cdots <_\mathrm{T} X$. By the relativized form of Theorem 9.42, every $X_i$-computable nontrivial family of sets has an $X_{i+1}$-computable maximal subfamily with the finite intersection property. Thus $\{Y : \exists i\, Y \leqslant_\mathrm{T} X_i\}$ is an $\omega$-model of $\mathrm{RCA}_0 + \mathrm{FIP}$ consisting entirely of $X$-computable sets.

Such models are enough to show that FIP does not imply $P$ for a wide range of principles $P$. For example, we can take $X$ above to be low. (The existence of a noncomputable low c.e. set was shown by Friedberg (see Soare [196]).) Then we immediately get nonimplication results from FIP for any principle that does not always have low solutions, such as COH or $\mathrm{SRT}_2^2$. As mentioned in Section 3.1, the only c.e. sets of PA degree are the complete ones, so by the upwards closure of PA degrees, the only c.e. sets that compute a completion of PA are the complete ones. Thus we also see that RCA$_0$ + FIP $\nvdash$ WKL$_0$.

Conversely, by Corollary 4.25, there is an $\omega$-model $\mathcal{M}$ of WKL$_0$ consisting entirely of sets of hyperimmune-free degree. It follows from Theorem 9.43 that $\mathcal{M} \nvDash$ FIP, so WKL$_0$ $\nvdash$ FIP. We have to be a bit careful here, though. Let $\mathcal{A}$ be a nontrivial family of sets as in Theorem 9.43. Then $\mathcal{A} \in S_M$ (where $S_M$ is the second order part of $\mathcal{M}$). Let $\mathcal{B} \in S_M$ be a subfamily of $\mathcal{A}$ with the finite intersection property. Then $\mathcal{B}$ is not maximal, because it has hyperimmune-free degree, which means that there is a subfamily $\mathcal{C}$ of $\mathcal{A}$ with the finite intersection property such that $\mathcal{B}$ is a subfamily of $\mathcal{C}$ but $\mathcal{C}$ is not a subfamily of $\mathcal{B}$. However, there is no reason such a $\mathcal{C}$ has to be in $S_M$. (In other words, we do not know a priori that counterexamples to the maximality of $\mathcal{B}$ live in $\mathcal{M}$, which raises the possibility that $\mathcal{B}$ may in fact be maximal *within the model* $\mathcal{M}$, in which case we would not have a proof that $\mathcal{M} \nvDash$ FIP.)

Fortunately, we can get around this problem easily. There must be a set $A \in \mathcal{C}$ that is not in $\mathcal{B}$. From $\mathcal{B}$, we can computably obtain another subfamily $\mathcal{D}$ of $\mathcal{A}$ whose elements are exactly the elements of $\mathcal{B}$ and $A$. Since $\mathcal{D}$ is a subfamily of $\mathcal{C}$, it has the finite intersection property. Furthermore, $\mathcal{B}$ is a subfamily of $\mathcal{D}$, but $\mathcal{D}$ is not a subfamily of $\mathcal{B}$. Finally, $\mathcal{D} \in S_M$, so we do indeed have $\mathcal{M} \nvDash$ FIP. (In Section 10.2, we will see an example of a principle where we have to deal with a similar issue, but cannot get around it with this kind of trick.)

Dzhafarov and Mummert [52] showed that FIP follows from RT$_2^2$, at least over RCA$_0$+I$\Sigma_2^0$. (We will see in the next section that this implication in fact holds over RCA$_0$.) Their proof proceeds via a principle inspired by the computability theoretic tradition of studying effective versions of forcing. These notions have played an important role in computability theory, and have also proved of interest to reverse mathematics. (See Jockusch [100] and Cholak, Dzhafarov, and Hirst [18] for computability theoretic versions of Cohen forcing and Mathias forcing, respectively.)

When we say that $D_0, D_1, \ldots$ are *uniformly* $\Pi_1^0$ *predicates on* $2^{<\mathbb{N}}$, we mean that we fix a $\Pi_1^0$ formula $\varphi(i, \sigma)$ (possibly with parameters) and say that $D_i(\sigma)$ holds iff $\varphi(i, \sigma)$ holds. We say that $D_i$ is *dense* if every binary string has an extension $\sigma$ such that $D_i(\sigma)$ holds. A sequence $G \in 2^{\mathbb{N}}$ *meets* $D_i$ if there is an initial segment $\sigma$ of $G$ such that $D_i(\sigma)$ holds.

**Definition 9.44.** $\Pi_1^0$G is the statement that for any collection of uniformly

$\Pi_1^0$ dense predicates $D_0, D_1, \ldots$ on $2^{<\mathbb{N}}$, there is a $G \in 2^{\mathbb{N}}$ that meets each $D_i$.

It is not difficult to show that $\Pi_1^0\mathsf{G}$ is computability theoretically equivalent to the statement that for any collection of uniformly $\Delta_2^0$ dense predicates $D_0, D_1, \ldots$ on $2^{<\mathbb{N}}$, there is a $G \in 2^{\mathbb{N}}$ that meets each $D_i$. In this form, this principle was considered by Shinoda and Slaman [178] and later by Csima, Hirschfeldt, Knight, and Soare [33] in connection with the study of atomic models, which we will discuss in the next section.

$\boxed{74}$ **Exercise 9.45** (Dzhafarov and Mummert [52]). Show that $\mathsf{RCA}_0 + \Pi_1^0\mathsf{G} \vdash \mathsf{FIP}$. [Hint: Given a nontrivial family of sets $A_0, A_1, \ldots$, consider the notion of forcing whose conditions are pairs $(\sigma, n)$, where $\sigma \in \mathbb{N}^{<\mathbb{N}}$ and $\bigcap_{i < |\sigma|} A_{\sigma(i)}$ contains some number $\leqslant s$, and where $(\tau, t)$ extends $(\sigma, s)$ if $\sigma$ is an initial segment of $\tau$ and $t \geqslant s$.]

Diamondstone, Downey, Greenberg, and Turetsky [37] clarified the connection between FIP and genericity by tying FIP to the important computability notion of 1-genericity. We will discuss some further details of the connections between FIP, $\Pi_1^0\mathsf{G}$, $\mathsf{RT}_2^2$, and other principles mentioned above in the next section.

## 9.3 Atomic models and omitting types

Another area of mathematics where principles provable from $\mathsf{RT}_2^2$ can be found is basic model theory. In this section, we will focus on atomic models and type omitting, but there is also considerable related material on the reverse mathematics of homogeneous models (see [32, 87, 120–122]). The common thread in much of this work is the combinatorics of omitting and realizing types. It is possible to abstract away the model theoretic framework of several theorems in this area to reveal a combinatorial core that connects them with principles of the kind we have been discussing. We will see examples of this process in this section.

We begin by reviewing a few definitions. See Chang and Kiesler [17] or Marker [132] for an introduction to model theory, and Ash and Knight [3] and Harizanov [77] for more on computable model theory. In this section, by a *theory* we mean a consistent set of sentences in a countable first order language, and all our structures are countable. Let $\mathcal{A}$ be a structure in a language $\mathcal{L}$. The *theory* of $\mathcal{A}$ is the complete theory consisting of all $\mathcal{L}$-sentences that hold in $\mathcal{A}$. Let $\mathcal{L}'$ be the expansion of $\mathcal{L}$ obtained by adding a constant for each element of $\mathcal{A}$. We think of $\mathcal{A}$ as a structure in the language $\mathcal{L}'$ in the obvious way. The *elementary diagram* of $\mathcal{A}$ is the set of all $\mathcal{L}'$-sentences that hold in $\mathcal{A}$. The *atomic diagram* of $\mathcal{A}$ is the set of all quantifier-free $\mathcal{L}'$-sentences that hold in $\mathcal{A}$.

Following traditional terminology, we call a computable theory *decidable*. A *decidable* structure is one whose elementary diagram is computable. Here there is a distinction made in computable model theory between a decidable structure and a *computable* structure, which is one whose atomic diagram is computable. Although both concepts are of great interest, the former is better suited to studying the strength of model theoretic theorems connecting theories with their models, as it matches up better with the notion of a decidable theory. (A decidable structure clearly has a decidable theory, but the theory of a computable structure might be as complex as $\emptyset^{(\omega)} = \{\langle n, i\rangle : i \in \emptyset^{(n)}\}$; this is the case for the natural numbers, for instance.) Similarly, for the purposes of reverse mathematics, we identify a structure with its elementary diagram. If instead we chose to identify a structure with its atomic diagram, even the statement that every structure has a theory would imply $\mathrm{ACA}_0$, which suggests that this choice would be the wrong one for our reverse mathematical analysis to yield meaningful information on model theoretic principles. For formal definitions in $\mathrm{RCA}_0$, see Section II.8 in Simpson [191]. That section also includes basic results such as a proof of the completeness theorem in $\mathrm{RCA}_0$.

A *partial type* of a theory $T$ is a set of formulas in a fixed tuple of free variables that is consistent with $T$. By a *type* of $T$ we will mean a complete type of $T$, i.e., a maximal partial type of $T$. A partial type $\Gamma$ is *realized* in a model $\mathcal{A}$ of $T$ if there is a tuple $\vec{a}$ of elements of $\mathcal{A}$ such that $\mathcal{A} \vDash \varphi(\vec{a})$ for every $\varphi \in \Gamma$. Otherwise, $\Gamma$ is *omitted* in $\mathcal{A}$.

A formula $\varphi(\vec{x})$ of $T$ is an *atom* of $T$ if for each formula $\psi(\vec{x})$ we have $T \vdash \varphi \rightarrow \psi$ or $T \vdash \varphi \rightarrow \neg\psi$, but not both. A partial type $\Gamma$ is *principal* if there is a formula $\varphi$ consistent with $T$ such that $T \vdash \varphi \rightarrow \psi$ for all $\psi \in \Gamma$. Thus a (complete) type is principal iff it contains an atom of $T$.

A theory $T$ is *atomic* if for every formula $\psi(\vec{x})$ consistent with $T$, there is an atom $\varphi(\vec{x})$ of $T$ extending it, i.e., one such that $T \vdash \varphi \rightarrow \psi$. A structure $\mathcal{A}$ is *atomic* if every tuple of elements of $\mathcal{A}$ satisfies an atom of its theory, i.e., if every type realized in $\mathcal{A}$ is principal. A structure $\mathcal{A}$ is *prime* if it can be elementarily embedded in every model of its theory.

Hirschfeldt, Shore, and Slaman [89] analyzed the strength of the following basic facts on atomic and prime models. (In these statements, it should be kept in mind that all theories and structures are countable.)

1. A complete theory has an atomic model iff it is atomic.
2. Any two atomic models of a given complete theory are isomorphic.
3. A complete theory has a prime model iff it is atomic.
4. Any two prime models of a given complete theory are isomorphic.
5. A structure is prime iff it is atomic.

It is easy to check that the standard proofs of the above theorems show that the "only if" directions of 1, 3, and 5 are provable in $\mathrm{RCA}_0$. Several other parts of these theorems are equivalent to $\mathrm{ACA}_0$, as we now see.

|75| **Exercise 9.46** (Hirschfeldt, Shore, and Slaman [89]). Let $f : \mathbb{N} \to \mathbb{N}$ be an injection. We work in a language with unary predicates $R_i$ and $R_{i,j}$ for $i, j \in \mathbb{N}$. Let $\mathcal{A}$ and $\mathcal{B}$ be two structures in this language, both with domain $\mathbb{N}$, defined as follows.

First, both $R_e^{\mathcal{A}}$ and $R_e^{\mathcal{B}}$ hold exactly of the numbers of the form $\langle e, n \rangle$. If $\forall t \leqslant s \, [f(t) \neq e]$ then both $R_{e,s}^{\mathcal{A}}$ and $R_{e,s}^{\mathcal{B}}$ are empty. Suppose $f(t) = e$ and $s \geqslant t$. Then $R_{e,s}^{\mathcal{A}}$ consists exactly of those numbers of the form $\langle e, n \rangle$ where either $n \leqslant t$ or $n$ is even; and $R_{e,s}^{\mathcal{B}}$ consists exactly of those numbers of the form $\langle e, n \rangle$ where $n > t$ and $n$ is even.

Arguing within $\mathrm{RCA}_0$, do the following:

a. Give a set of axioms $S$ that holds in both $\mathcal{A}$ and $\mathcal{B}$ and admits effective quantifier elimination; i.e., there is a function taking each formula $\psi$ to a quantifier-free formula $\varphi$ (which could be just one of the formal propositional symbols T, for true, or F, for false) with the same free variables as $\varphi$ such that $S \vdash \varphi \leftrightarrow \psi$. Conclude that $S$ has a deductive closure $T$ that is a complete theory, and that $\mathcal{A}$ and $\mathcal{B}$ are structures (i.e., the elementary diagrams of these structures exist). [See e.g. Enderton [53] for more on quantifier elimination, in particular the fact that one can restrict attention to those formulas $\psi$ of the form $\exists x \, \theta$, where $\theta$ is a conjunction of atomic formulas and negations of atomic formulas.]

b. Show that $T$ is an atomic theory and $\mathcal{A}$ and $\mathcal{B}$ are atomic models of $T$.

c. Show that if $\mathcal{C} \vDash T$ and there are embeddings $\mathcal{C} \to \mathcal{A}$ and $\mathcal{C} \to \mathcal{B}$, then the range of $f$ exists.

d. Use the above to show that the following statements are equivalent to $\mathrm{ACA}_0$ over $\mathrm{RCA}_0$:

   (a) Every atomic structure is prime.
   (b) Every complete atomic theory has a prime model.
   (c) Any two atomic models of a given complete theory are isomorphic.

The above results show that $ACA_0$ proves that any two prime models of a given complete theory are isomorphic, but the following question is otherwise open.

▶ **Open Question 9.47.** What is the reverse mathematical strength of the statement that any two prime models of a given complete theory are isomorphic?

We are thus left with the following principle.

**Definition 9.48.** The *Atomic Model Theorem* (AMT) is the statement that every complete atomic theory has an atomic model.

The computability theoretic analysis of AMT has its roots in the early days of computable model theory. If a theory has countably many types then it is atomic (see e.g. Marker [132] for a proof of this fact), and hence has an atomic model. The following theorem shows that this fact holds effectively; it is a consequence of Theorem 9.63 below. (Millar [141] was working in the computability theoretic context, but his proof carries through in $RCA_0$.) By a *listing* of the types of $T$ we mean a sequence $p_0, p_1, \ldots$ of types of $T$ that includes every type of $T$. We do allow repetitions (i.e., we could have $p_i = p_j$ for $i \neq j$).

**Theorem 9.49** (Millar [141]). *The following is provable in* $RCA_0$. *Let $T$ be a complete theory such that there is a listing of the types of $T$. Then $T$ has an atomic model.*

In contrast, we have the following theorem, which we will prove below.

**Theorem 9.50** (Goncharov and Nurtazin [72]; Millar [140]). *There is a decidable theory $T$ such that each type of $T$ is computable but $T$ has no decidable (or even computable) atomic model.*

In particular, $RCA_0 \nvdash AMT$. Theorem 9.50 was extended as follows (without the assumption that each type of $T$ is computable; we will see in Theorem 9.70 below that there is no possible extension with that assumption). Say that a set $X$ is *atomic bounding* if every complete decidable atomic theory has an $X$-decidable atomic model.

**Theorem 9.51** (Csima, Hirschfeldt, Knight, and Soare [33]). *A $\Delta_2^0$ set is atomic bounding iff it is not low$_2$.*

Outside of the $\Delta_2^0$ degrees, the situation is more complex; see Conidis [28].

Let $Y$ be a set of PA degree and let $\mathcal{M}$ be an $\omega$-model of $WKL_0$ all of whose sets are $Y$-computable, which exists by Theorem 4.22. Then Theorem 9.51 implies that $\mathcal{M} \nvDash AMT$, so $WKL_0 \nvdash AMT$. Nevertheless, AMT is quite a weak principle, as the following theorems show.

**Theorem 9.52** (Hirschfeldt, Shore, and Slaman [89]). $RCA_0 + SADS \vdash$ AMT.

**Theorem 9.53** (Hirschfeldt, Shore, and Slaman [89]). $RCA_0 + AMT$ *is* $r$-$\Pi_2^1$-*conservative over* $RCA_0$. *In particular,* $RCA_0 + AMT \nvdash SADS$ *and* $RCA_0 + AMT \nvdash WKL_0$.

We will discuss the proofs of these theorems below. As mentioned in Section 9.1, $RCA_0 + SADS$ has an $\omega$-model consisting entirely of low sets, so the following is one consequence of Theorem 9.52, originally proved directly. (We will see another proof in Exercise 9.59 below.)

**Theorem 9.54** (Csima [31]). *Every complete decidable atomic theory has an atomic model whose elementary diagram is low.*

The analysis of AMT provides a good example of the practice of stripping away the area-specific details of a theorem to reveal its combinatorial core. The first step is to get rid of models.

**Theorem 9.55** (Hirschfeldt, Shore, and Slaman [89]). *The following is provable in* $RCA_0$. *Let* $T$ *be a complete atomic theory such that there is a listing of the principal types of* $T$. *Then* $T$ *has an atomic model.*

Before giving the details of the proof of this theorem, let us discuss an important technique that it illustrates. The proof of Theorem 9.55 is based on that of its earlier computability theoretic analog.

**Theorem 9.56** (Goncharov and Nurtazin [72]; Harrington [79]). *Let* $T$ *be a complete decidable atomic theory such that there is a computable listing of the principal types of* $T$. *Then* $T$ *has a decidable atomic model.*

The proof of Theorem 9.56 uses a Henkin construction (as in the usual proof of the completeness theorem, e.g. in Enderton [53]) to build a model $\mathcal{A}$, with additional requirements ensuring that each tuple $\vec{a}$ of elements of $\mathcal{A}$ satisfies some atom of $T$. These requirements would be easy to satisfy if we knew the atoms of $T$, but, in general, we cannot obtain atoms of $T$ effectively from the set of principal types of $T$. (See the last part of Exercise 9.57 below.) So we use our computable listing of principal types to assign a principal type $p_{\vec{a}}$ to each $\vec{a}$. We then proceed with our Henkin construction, attempting to ensure that each $\vec{a}$ has type $p_{\vec{a}}$. Of course, it may not be consistent for tuples $\vec{a}$ and $\vec{b}$ to have the types $p_{\vec{a}}$ and $p_{\vec{b}}$, respectively. If we find that to be the case, we have to assign a new type to one of these tuples. We handle this situation by ordering the tuples of elements of $\mathcal{A}$ into a priority list. In the above situation, we assign a new type to the tuple of weaker priority (i.e., the one lower down on the list), saying that this tuple is injured by the higher priority one. The construction is thus a priority argument. It is a finite injury argument because as we make $\vec{a}$ satisfy more and more of the type $p_{\vec{a}}$, we eventually ensure that $\vec{a}$ satisfies an atom. After that point, $\vec{a}$ can no longer cause any injuries. See Goncharov and Nurtazin [72], Harrington [79], or Hirschfeldt, Shore, and Slaman [89] for details.

A finite injury argument is usually split into a construction and a verification that the construction succeeds. Typically, the construction can be carried out in $\mathrm{RCA}_0$. The verification, however, is usually an inductive argument, in which one shows that a given requirement is eventually satisfied and stops acting by assuming that each stronger priority requirement eventually stops acting. The statement that a requirement $R$ eventually stops acting can be restated as "there is a stage $s$ such that for all stages $t \geqslant s$, the requirement $R$ does not act at stage $t$." This is a $\Sigma_2^0$ statement, so a priori we are dealing with $\Sigma_2^0$-induction, which is not provable in $\mathrm{RCA}_0$. Indeed, there are finite injury arguments that cannot be carried out in $\mathrm{RCA}_0$, and the study of the proof theoretic strength of priority arguments and other computability theoretic constructions has been a fruitful line of research, known as reverse computability theory (or, more commonly, reverse recursion theory); see Chong and Yang [25, 26].

In some cases, however, we can modify a priority construction to eliminate the need for $\Sigma_2^0$-induction. One technique for doing so was originally developed in the context of the theory of computability on admissible sets known as $\alpha$-recursion theory by Shore [181], and was later called *Shore blocking* in applications to reverse computability theory in structures in the language of first order arithmetic (as in Mytilinaios [153]). This technique involves creating blocks of requirements so that there is no injury within each block. Model theory turns out to be a natural setting for blocking arguments. For example, suppose we have a block of requirements $R_0, \ldots, R_n$, where $R_i$ says that the type realized by a tuple $\vec{a}_i$ in a model we are building is principal. Then we can replace all these requirements by one requirement stating that the type realized by the tuple $\vec{a}_0 \vec{a}_1 \ldots \vec{a}_n$ is principal. Satisfying this "super-requirement" ensures that all of $R_0, \ldots, R_n$ are satisfied, since subtypes of principal types are principal. The proof of Theorem 9.55 is a relatively simple illustration of this method.

*Proof of Theorem 9.55.* We argue in $\mathrm{RCA}_0$. Let $\mathcal{L}$ be the language obtained by adding new constant symbols $c_0, c_1, \ldots$ to the language of $T$, and let $\varphi_0, \varphi_1, \ldots$ be a listing of all the sentences of $\mathcal{L}$. We define a tree $\mathcal{S}$, representing possible Henkin constructions of models of $T$, by recursion. Each node $\sigma \in \mathcal{S}$ will be labeled by a set $S_\sigma$ of sentences of $\mathcal{L}$ consistent with $T$. Begin by putting the empty string $\lambda$ into $\mathcal{S}$ and letting $S_\lambda = \emptyset$. Now suppose we have put $\sigma$ of length $n$ into $\mathcal{S}$. First, if $\varphi_n$ is consistent with $T \cup S_\sigma$, then put $\sigma 1$ into $\mathcal{S}$. In this case, if $\varphi_n$ is of the form $\exists x \psi(x)$, choose the least $i$ such that $c_i$ does not appear in $S_\sigma$ and let $S_{\sigma 1} = S_\sigma \cup \{\varphi_n, \psi(c_i)\}$; otherwise let $S_{\sigma 1} = S_\sigma \cup \{\varphi_n\}$. Next, if $\neg \varphi_n$ is consistent with $T \cup S_\sigma$, then put $\sigma 0$ into $\mathcal{S}$ and let $S_{\sigma 0} = S_\sigma \cup \{\neg \varphi_n\}$.

The usual arguments show that $\mathcal{S}$ has no dead ends and that if $\sigma i \in \mathcal{S}$, then $S_{\sigma i}$ is consistent with $T$. If $P$ is a path on $\mathcal{S}$, then $S_P = \bigcup_{\sigma \in P} S_\sigma$ is consistent (and contains $T$) and can be used to define a model $\mathcal{M}_p$ of $T$ by the usual Henkin argument. (It is straightforward to verify that all of these arguments can be carried out in $\mathrm{RCA}_0$.)

Let $\Gamma_0, \Gamma_1, \ldots$ be the principal types of $T$. For $\sigma \in \mathcal{S}$ and $n \in \mathbb{N}$, let $t(\sigma, n)$ be the least $t$ such that $\exists x_{n+1}, \ldots, x_m \bigwedge_{\varphi \in S_\sigma} \varphi[c_0/x_0, \ldots, c_m/x_m] \in \Gamma_t$, where $c_m$ is the largest constant mentioned in $S_\sigma$ (and $c/x$ means that $c$ is substituted by $x$). For a path $P$ on $\mathcal{S}$, let $t(P, n) = \lim_k t(P \restriction k, n)$, which of course can be infinite. Note that if $t(P, n) < \infty$, then $c_0, \ldots, c_n$ has type $\Gamma_{t(P,n)}$ in $\mathcal{M}_P$. Thus, if $t(P, n) < \infty$ for all $n$, then $\mathcal{M}_P$ is an atomic model of $T$.

So we see that our task is to define a path $P$ on $\mathcal{S}$ while satisfying the requirements

$$R_n : \exists j \, \forall k \geqslant j \, [t(P \restriction j, n) = t(P \restriction k, n)].$$

As discussed above, if we try to do so with a standard priority construction, we will run into the difficulty of not being able to verify by induction that all requirements are satisfied, since we lack $\mathrm{I}\Sigma_2^0$. So at each stage $k$ of our construction, we have numbers $n_0, n_1, \ldots$ specifying *blocks* of requirements. These numbers can increase during the construction. The $i$th block of requirements consists of all $R_n$ with $n \leqslant n_i$. (It does not matter here that the blocks overlap.)

The key observation is the following: Suppose we ensure that $R_m$ holds for some $m > n$, so that $t(P, m)$ has a finite value $e$. Then the tuple

$c_0, \dots, c_m$ has type $\Gamma_e$, and hence satisfies some atom $\alpha$ of $T$. But then the tuple $c_0, \dots, c_n$ satisfies $\beta \equiv \exists x_{n+1}, \dots, x_m\, \alpha$, which is also an atom of $T$. Thus $t(P, n)$ is the least $i$ such that $\Gamma_i$ is an $(n+1)$-type containing $\beta$, and hence $R_n$ holds.

So at stage $k$, we define the $k$th bit of $P$ in such a way that $t(P \upharpoonright k + 1, n_i) = t(P \upharpoonright k, n_i)$ for as large an $i$ as possible. Of course, this action may cause an injury to a lower priority block (i.e., the $j$th block for some $j > i$). To handle this issue, we redefine each $n_j$ for $j \geqslant i$ to be at least $k$. We will then be able to argue, using only $\Sigma_1^0$-induction, that for each $n$ there is an $i$ such that $n_i$ eventually settles on a value greater than $n$. It must then be the case that the $i$th block eventually stops being injured, and hence becomes in effect the highest priority block, which ensures that $R_{n_i}$ is satisfied, and hence so is $R_n$.

We now proceed with the construction and verification. The value of $n_i$ at stage $k$ will be denoted by $n_{i,k}$. Let $n_{i,0} = i$ for all $i$. At stage $k$, if $\sigma = P \upharpoonright k$ has a single successor $\sigma j$ in $S$ then let $P(k) = j$ and let $n_{i,k+1} = n_{i,k}$ for all $i$. Note that in this case $t(P \upharpoonright k + 1, n) = t(P \upharpoonright k, n)$ for all $n$.

If $\sigma$ has two successors, then for $j \in \{0, 1\}$, let $d_{\sigma j}$ be the least $i \leqslant k$ such that $t(\sigma j, n_{i,k}) > t(\sigma, n_{i,k})$, or $d_j = \infty$ if there is no such $i$. If $d_{\sigma 0} = d_{\sigma 1}$ then let $j = 0$. (Note that this case can happen only if $d_{\sigma 0} = d_{\sigma 1} = \infty$.) Otherwise, let $j$ be such that $d_{\sigma j} > d_{\sigma(1-j)}$. Let $P(k) = j$. For $i < d_{\sigma j}$, let $n_{i,k+1} = n_{i,k}$. For $i \geqslant d_{\sigma j}$, let $n_{i,k+1} = k + i$.

Having defined $P$, let $n_i = \lim_k n_{i,k}$, which can be infinite. We claim that the finite values of $n_i$ are unbounded. In other words, for each $n$ there is an $i$ such that $n < n_i < \infty$.

Suppose otherwise, and let $n$ be such that every $n_i$ is either less than $n$ or infinite. Let $e$ be least such that $n_{e,k} > n$ for some $k$. Such an $e$ exists by $\mathrm{I}\Sigma_1^0$ (and the fact that there exist $i, k$ such that $n_{i,k} > n$). If $i < e$, then $i < d(P \upharpoonright k + 1, n_i)$ for all $k \geqslant n$, so $t(P \upharpoonright k + 1, n_i) = t(P \upharpoonright k, n_i)$ for all $k \geqslant n$, and hence $t(P, n_i) < \infty$. If $e > 0$ then let $\alpha$ be an atom of $T$ in $\Gamma_{t(P, n_{e-1})}$ and for each $i < e$, let $\alpha_i \equiv \exists x_{n_i + 1}, \dots, x_{n_{e-1}}\, \alpha$. Notice that each $\alpha_i$ is an atom of $T$, and each $\alpha_i[x_0/c_0, \dots, x_{n_i}/c_{n_i}]$ is in $P$. Let $m$ be such that $S_{P \upharpoonright m}$ contains $\alpha_i[x_0/c_0, \dots, x_{n_i}/c_{n_i}]$ for all $i < e$, or $m = 0$ if $e = 0$. Then for all extensions $\tau$ of $P \upharpoonright m$ in $S$ and all $i < e$, we must have $t(\tau, n_i) = t(P \upharpoonright m, n_i) = t(P, n_i)$.

Since $n_e > n$, we have $n_e = \infty$, so there is an $l \geqslant n$ such that both successors of $P \upharpoonright l$ are in $S$ and $d_{P \upharpoonright l+1} \leqslant e$. Let $\tau$ be the successor of $P \upharpoonright l$ other than $P \upharpoonright l + 1$. By the construction of $P$ we have $d_\tau < d_{P \upharpoonright l+1} \leqslant e$, so for some $i < e$ we have $t(\tau, n_i) > t(P \restriction l, n_i) = t(P, n_i)$, contradicting the last statement of the previous paragraph.

Thus the finite values of $n_i$ are unbounded. So given $n$ there is an $i$ such $n < n_i < \infty$. For all sufficiently large $k$ we must have $d_{P \restriction k} > i$, and hence $t(P \restriction k + 1, n_i) = t(P \restriction k, n_i)$, which implies that $t(P \restriction k + 1, n) = t(P \restriction k, n)$. So $t(P, n) < \infty$. Thus $\mathcal{M}_P$ is an atomic model of $T$.                                                        $\square$

Further examples of the use of Shore blocking in the reverse mathematics of model theory can be found in Hirschfeldt, Shore, and Slaman [89] and Hirschfeldt, Lange, and Shore [87]. One of these examples is the proof of Theorem 9.52, which we now discuss. It is easy to see that the existence of a listing of the principal types of a theory $T$ is equivalent to the existence of a procedure $P$ taking each formula $\varphi$ consistent with $T$ to a principal type of $T$ containing $\varphi$. Of course, if we have an oracle $X$ with sufficient power to compute a function taking each such $\varphi$ to an atom extending $\varphi$, then there is such a $P$ that is $X$-computable. However, as we will see in the last part of Exercise 9.57, in general the weakest oracle with this much power is $\emptyset'$.

We can get by with much less oracle power. In defining the type $P(\varphi)$, we can keep guessing at atoms, and will succeed in making $P(\varphi)$ principal as long as one such guess is correct. The proof of Theorem 9.52 proceeds by constructing a linear order $\preccurlyeq$ of type $\omega + \omega^*$ such that any infinite ascending or descending sequence has enough oracle power to make such eventually correct guesses. The way we code information into these sequences is by creating large gaps in each side of $\preccurlyeq$. For example, if we ensure that every number $i$ such that $m \leqslant i \leqslant n$ (where $\leqslant$ is the usual order on the natural numbers) is in the $\omega^*$ part of $\preccurlyeq$, then the $(m + 1)$st element of any infinite ascending sequence in $\preccurlyeq$ must be greater than $n$. So we need to create large enough gaps on each side to encode the guesses mentioned above. Of course, to create gaps in the $\omega$ side we need to put numbers into the $\omega^*$ side, and vice-versa, so there is a tension between encoding information into ascending sequences and into descending sequences, which we can resolve by using a priority argument. To make this argument carry through in $\mathrm{RCA}_0$ requires Shore blocking. See [89] for details.

Having removed the models from the statement of AMT, we can now remove the remaining model theoretic notions as follows. Recall the following definitions from Section 4.6. Let $\mathcal{T}$ be a binary tree with no dead ends. A node $\sigma \in T$ is an atom if for each $n > |\sigma|$, there is exactly one extension of $\sigma$ of length $n$ in $\mathcal{T}$. The tree $\mathcal{T}$ is atomic if each node of $\mathcal{T}$ can be extended to an atom, and strongly atomic if for every $\sigma_0, \ldots, \sigma_n \in \mathcal{T}$, each $\sigma_i$ can be extended to an atom. As shown in Exercise 4.35, the equivalence between these two notions cannot be established in $\text{RCA}_0$, but of course the two notions are equivalent in any $\omega$-model. A path on $\mathcal{T}$ is *isolated* if it contains an isolating node.

### 76   Exercise 9.57.

**a.** Let $T$ be a complete decidable atomic theory. Show that there is a computable atomic binary tree $\mathcal{T}$ with no dead ends such that the set of degrees of paths on $\mathcal{T}$ is the same as the set of degrees of types of $T$, and such that any listing of the isolated paths on $\mathcal{T}$ computes a listing of the principal types of $T$.

**b.** Let $\mathcal{T}$ be a computable atomic binary tree with no dead ends. Show that there is a complete decidable atomic theory $T$ that admits effective quantifier elimination, such that the set of degrees of types of $T$ is the same as the set of degrees of paths on $\mathcal{T}$, and such that any listing of the principal types of $T$ computes a listing of the isolated paths on $\mathcal{T}$.

**c.** Adapt the constructions in the previous two parts to show that AMT is equivalent over $\text{RCA}_0$ to the statement that for any strongly atomic binary tree $\mathcal{T}$ with no dead ends, there is a listing of the isolated paths on $\mathcal{T}$.

**d.** Use the construction in part b to show that there is a complete decidable theory $T$ such that there is a computable listing of the principal types of $T$, but any function taking each formula $\varphi$ consistent with $T$ to an atom extending $\varphi$ computes $\emptyset'$.

Thus the following result has Theorem 9.50 as a corollary (obtained by applying the second part of the above exercise, and noting that the fact that the theory $T$ admits effective quantifier elimination implies that any computable model of $T$ is in fact decidable).

**Theorem 9.58** (Goncharov and Nurtazin [72]). *There is a computable binary tree $\mathcal{T}$ with no dead ends such that each path on $\mathcal{T}$ is computable but there is no computable listing of the isolated paths on $\mathcal{T}$.*

*Proof.* Let $\Psi_0, \Psi_1, \ldots$ be an effective list of all partial computable functions from $\mathbb{N}^2$ to $\{0, 1\}$. Let $\Psi^n_e(x) = \Psi_e(n, x)$. It is enough to satisfy each requirement $R_e$ stating that either there is an isolated path on $\mathcal{T}$ that is not equal to $\Psi^n_e$ for any $n$, or there is an $n$ such that $\Psi^n_e$ is not an isolated path on $\mathcal{T}$. Begin by ensuring that $1^e 0 \in \mathcal{T}$ for all $e$. We use the part of $\mathcal{T}$

above $1^e0$ to satisfy $R_e$.

For each $e$, proceed as follows. Start building an isolated path above $1^e0$ until we find some $n$ such that $\Psi_e^n \upharpoonright e + 1 = 1^e0$. If there is no such $n$ then the isolated path above $1^e0$ is not equal to $\Psi_e^n$ for any $n$, so $R_e$ is satisfied. If $n$ is found then introduce a splitting above $1^e0$ (i.e., ensure that there is a $\sigma$ extending $1^e0$ such that $\sigma0$ and $\sigma1$ are both in $\mathcal{T}$). Continue to build isolated paths above $\sigma0$ and $\sigma1$, until $\Psi_e^n$ is seen to extend $\sigma i$ for some $i \in \{0, 1\}$. If this event never occurs then $\Psi_e^n$ is not a path on $\mathcal{T}$, so $R_e$ is satisfied. Otherwise, introduce another splitting above $\sigma i$, and wait until $\Psi_e^n$ is seen to extend one of the sides of this splitting. Continue to build $\mathcal{T}$ in this way. If $\Psi_e^n$ is a path on $\mathcal{T}$, then this procedure ensures that there are infinitely many nodes $\sigma$ in $\Psi_e^n$ such that both $\sigma0$ and $\sigma1$ are in $\mathcal{T}$, and hence (since $\mathcal{T}$ clearly has no dead ends) $\Psi_e^n$ is not isolated.

It is straightforward to formalize this construction to see that $\mathcal{T}$ can be built computably. Each path on $\mathcal{T}$ (other than the path consisting of all 0's) is either isolated, and hence computable, or equal to $\Psi_e^n$ for some $e$ and $n$, and hence also computable.                                                               □

Recall the principle $\Pi_1^0 G$ from Definition 9.44. Being an atom of a computable tree with no dead ends is a $\Pi_1^0$ property (as is being an atom of a decidable theory), so the following fact should not be surprising.

**77** **Exercise 9.59** (Csima, Hirschfeldt, Knight, and Soare [33]; Hirschfeldt, Shore, and Slaman [89]; Csima [31]).
**a.** Show that $\mathrm{RCA}_0 + \Pi_1^0 G \vdash \mathrm{AMT}$.
**b.** Use this fact to prove Theorem 9.54.

In fact, the combinatorics of AMT are almost the same as those of $\Pi_1^0 G$. Conidis [28] showed that $\Pi_1^0 G \leqslant_c \mathrm{AMT}$, so in particular every $\omega$-model of $\mathrm{RCA}_0 + \mathrm{AMT}$ is a model of $\Pi_1^0 G$. As noted in [89], a straightforward analysis of his proof yields the following result.

**Theorem 9.60** (Conidis [28]). $\mathrm{RCA}_0 + \mathrm{I}\Sigma_2^0 + \mathrm{AMT} \vdash \Pi_1^0 G$.

AMT and $\Pi_1^0 G$ have interesting interactions with first order principles.

**78** **Exercise 9.61** (Hirschfeldt, Shore, and Slaman [89]). Show that $\mathrm{RCA}_0 + \Pi_1^0 G$ is r-$\Pi_2^1$-conservative over $\mathrm{RCA}_0$. [Hint: Adapt the proof of Theorem 7.18 to Cohen forcing in place of Mathias forcing.]

Combining the solution to this exercise with that of Exercise 9.59 yields a proof of Theorem 9.53. Hirschfeldt, Shore, and Slaman [89] also showed that AMT is $\Pi_1^1$-conservative over $\mathrm{B}\Sigma_2^0$ and r-$\Pi_2^1$-conservative over $\mathrm{I}\Sigma_2^0$ (cf. the discussion of WKL and COH at the end of Section 7.2), and that $\Pi_1^0\mathrm{G}$ is r-$\Pi_2^1$-conservative over $\mathrm{I}\Sigma_2^0$. (Conservativity over $\mathrm{I}\Sigma_2^0$ comes from the same proof as conservativity over $\mathrm{RCA}_0$, because adding Cohen generics preserves $\mathrm{I}\Sigma_2^0$ as well as $\mathrm{I}\Sigma_1^0$.) On the other hand, we have the following result.

**Theorem 9.62** (Hirschfeldt, Shore, and Slaman [89]). $\mathrm{RCA}_0 + \mathrm{B}\Sigma_2^0 + \Pi_1^0\mathrm{G} \vdash \mathrm{I}\Sigma_2^0$.

One consequence of the above results is that AMT does not imply $\Pi_1^0\mathrm{G}$ over $\mathrm{RCA}_0$, or even over $\mathrm{RCA}_0 + \mathrm{B}\Sigma_2^0$. Another is that $\mathrm{RCA}_0 + \mathrm{RT}_2^2 \nvdash \Pi_1^0\mathrm{G}$, since $\mathrm{RCA}_0 + \mathrm{RT}_2^2 \vdash \mathrm{B}\Sigma_2^0$, by Theorems 6.81 and 6.82, but $\mathrm{RCA}_0 + \mathrm{RT}_2^2 \nvdash \mathrm{I}\Sigma_2^0$, by Theorem 6.85. For more on the first order consequences of $\Pi_1^0\mathrm{G}$ and related principles, see Hirschfeldt, Lange, and Shore [87].

We now turn to two other principles related to AMT. The classical omitting types theorem says that for any countable set $S$ of nonprincipal types of a complete theory $T$, there is a model of $T$ that omits all types in $S$. This form of the omitting types theorem holds in $\mathrm{RCA}_0$. Indeed, we have the following stronger result; it was proved by Millar [141] in a computability theoretic version, but as noted in [89], his proof carries through in $\mathrm{RCA}_0$.

**Theorem 9.63** (Millar [141]). *The following is provable in* $\mathrm{RCA}_0$. *Let* $T$ *be a complete theory, let* $S_0$ *be a set of (complete) types of* $T$, *and let* $S_1$ *be a set of nonprincipal partial types of* $T$. *Then there is a model of* $T$ *omitting all nonprincipal types in* $S_0$ *and all partial types in* $S_1$.

One corollary to this result is Theorem 9.49, which points to a tie between the effectiveness of atomic models of a complete atomic theory $T$ and that of the types of $T$ (including the nonprincipal ones), a topic to which we will return below.

There *is* a version of the omitting types theorem that is not provable in $\mathrm{RCA}_0$.

**Definition 9.64.** The *Omitting Partial Types* (OPT) principle is the statement that for any complete theory $T$ and any set $S$ of partial types of $T$, there is a model of $T$ omitting all nonprincipal partial types in $S$.

Millar [141] showed that OPT is not computably valid, and in particular does not hold in the minimal $\omega$-model of $\mathrm{RCA}_0$. One way to establish this result is to combine Theorem 9.50 with the following fact.

**79** **Exercise 9.65.** Let $T$ be a complete decidable theory each of whose types is computable. Show that there is a computable set $S$ of partial types of $T$ such that any model omitting all nonprincipal partial types in $S$ is atomic. [Hint: We can transform a c.e. set of formulas into an equivalent computable set by noting that any formula $\varphi$ is equivalent to $\varphi \wedge \cdots \wedge \varphi$.]

As mentioned in Chapter 5, OPT is an example of a principle from outside computability theory whose strength is exactly captured by a computability theoretic notion, in this case hyperimmunity.

**Theorem 9.66** (Hirschfeldt, Shore, and Slaman [89]). 1. *Let $T$ be a complete decidable theory, let $S$ be a computable set of partial types of $T$, and let $X$ have hyperimmune degree. Then there is an $X$-decidable model of $T$ that omits all nonprincipal partial types in $S$.*

2. *There are a complete decidable atomic theory $T$ and a computable set $S$ of partial types of $T$ such that the atomic diagram of any model of $T$ that omits all nonprincipal partial types in $S$ has hyperimmune degree.*

The proof of this result can be adapted to show the following. Recall that a function $f$ is dominated by a function $g$ if $f(n) \leqslant g(n)$ for almost all $n$.

**Theorem 9.67** (Hirschfeldt, Shore, and Slaman [89]). OPT *is equivalent over* $\mathrm{RCA}_0$ *to the statement that for any set $X$, there is a set that has hyperimmune degree relative to $X$ (or, equivalently, there exists a function that is not dominated by any $X$-computable function).*

Since the theory in part 2 of Theorem 9.66 is atomic, the reverse mathematical version of that result has the following consequence.

**Theorem 9.68.** $\mathrm{RCA}_0 + \mathrm{AMT} \vdash \mathrm{OPT}$.

Recall the principle FIP from Definition 9.40. Theorem 9.43 can be adapted to show that $\mathrm{RCA}_0 + \mathrm{FIP}$ proves that for any set $X$, there is a set that has hyperimmune degree relative to $X$, so we have the following result.

**Corollary 9.69** (Dzhafarov and Mummert [52]). $\mathrm{RCA}_0 + \mathrm{FIP} \vdash \mathrm{OPT}$.

Of course, the comments following Theorem 9.43 on "small" $\omega$-models of FIP and the resulting nonimplication theorems hold for OPT as well. Indeed, OPT is even weaker than FIP. Recently, Diamondstone, Downey, Greenberg, and Turetsky [37] showed that $\mathrm{RCA}_0 + \mathrm{OPT} \nvdash \mathrm{FIP}$ (and indeed, there is an $\omega$-model of $\mathrm{RCA}_0 + \mathrm{OPT}$ that is not a model of FIP). A further clarification of the strength of FIP was provided by Day, Dzhafarov, and Miller [unpublished] and independently Greenberg and Hirschfeldt [unpublished], who showed that $\mathrm{RCA}_0 + \mathrm{AMT} \vdash \mathrm{FIP}$ (cf. Exercise 9.45).

The computability theoretic version of Theorem 9.49 is that every complete decidable theory whose types can be computably listed has an atomic model. Theorem 9.50 shows that the uniform computability provided by the listing in this statement is necessary. Nevertheless, the assumption that each type of a complete theory $T$ is computable is still quite strong, and makes it possible to find atomic models of $T$ "almost effectively", in the following sense.

**Theorem 9.70** (Hirschfeldt [84]). *Let $T$ be a complete decidable theory each of whose types is computable, and let $X$ be a noncomputable set. Then $T$ has an $X$-decidable atomic model.*

*Proof.* By Exercise 9.57, it is enough to take a computable binary tree $\mathcal{T}$ with no dead ends, all of whose paths are computable, and produce an $X$-computable listing of the isolated paths of $\mathcal{T}$. Let $\sigma_0, \sigma_1, \ldots$ be the nodes of $\mathcal{T}$. Let $P_n$ be the path on $\mathcal{T}$ defined as follows. Extend $\sigma_n$ until we find a split, i.e., a $\tau_0$ extending $\sigma_n$ such that both $\tau_0 0$ and $\tau_0 1$ are in $\mathcal{T}$, if ever. Follow $\tau_0 0$ if $0 \notin X$, and $\tau_0 1$ if $0 \in X$. Continue until we find another split $\tau_1 0, \tau_1 1$, if ever. Follow $\tau_1 0$ if $1 \notin X$, and $\tau_1 1$ if $1 \in X$. Continue in this manner, using successive bits of $X$ to decide which direction to take whenever a split is found.

The listing $P_0, P_1, \ldots$ is $X$-computable. If $P$ is an isolated path on $\mathcal{T}$ then there is an isolating node $\sigma \in P$. There is an $n$ such that $\sigma = \sigma_n$, and we must have $P_n = P$. Thus every isolated path on $\mathcal{T}$ is on our listing, so it is enough to show that each $P_n$ is isolated. Assume for a contradiction that $P_n$ is not isolated. Then in the above definition of $P_n$, we encounter infinitely many splits. Let $\tau_0, \tau_1, \ldots$ be as in that definition. For each $i$, we have $\tau_i 1 \in P_n$ iff $i \in X$. Thus we can compute $X$ from $P_n$. But $P_n$ is a path on $\mathcal{T}$, and hence is computable, contradicting the noncomputability of $X$. $\qquad \square$

This result is connected with other examples in computability theory and computable model theory in which certain noncomputable tasks can be performed by any noncomputable oracle; see [84] for more on these connections.

Theorem 9.70 has a reverse mathematical version. Partial types $\Gamma$ and $\Delta$ of a theory $T$ are *equivalent* if they imply the same formulas over $T$. A listing of partial types $\Gamma_0, \Gamma_1, \dots$ is a *subenumeration* of the types of a theory $T$ if for every (complete) type $p$ of $T$ there is an $i$ such that $\Gamma_i$ is equivalent to $p$. If the types of $T$ have a subenumeration then we say that they are *subenumerable*.

**Definition 9.71.** The *Atomic Model Theorem with Subenumerable Types* (AST) is the statement that every complete theory whose types are subenumerable has an atomic model.

The proof of Theorem 9.70 can easily be adapted to yield the following result, which shows that, like OPT, AST is equivalent to a natural computability theoretic principle, but in this case an even more basic one, namely the existence of noncomputable sets.

**Theorem 9.72** (Hirschfeldt, Shore, and Slaman [89]). AST *is equivalent over* $\mathrm{RCA}_0$ *to the statement that for any set $X$, there is a set $Y \not\leq_T X$.*

It is fair to say that this result shows that AST is the weakest natural principle that is not computably valid. Indeed, AST follows from every principle considered in this book, except for ones like $\mathrm{I}\Sigma_n^0$ that hold in all $\omega$-models, since for each such principle $P$, the proof that $P$ does not hold in the minimal $\omega$-model of $\mathrm{RCA}_0$ also shows that $P$ implies that for any set $X$, there is a set $Y \not\leq_T X$. In particular, AST is sufficiently weak to follow from both $\mathrm{WKL}_0$ and OPT. (The fact that $\mathrm{RCA}_0 + \mathrm{OPT} \vdash \mathrm{AST}$ can also be proved in a much more direct manner by adapting Exercise 9.65 to the reverse mathematical setting.)

While on the topic of model theoretic principles, it is worth mentioning an open question that, while not directly related to the above material, seems quite interesting. Let $\mathcal{A}$ be a structure, and let $T$ be the elementary diagram of $\mathcal{A}$, thought of as a complete theory in the language of $\mathcal{A}$ expanded by adding constants for the elements of $\mathcal{A}$. Recall that $\mathcal{A}$ is *saturated* if every type of $T$ is realized in $\mathcal{A}$ (where we think of $\mathcal{A}$ as a structure in the expanded language in the obvious way). Every complete theory with countably many types has a saturated model (see e.g. Marker [132] for a proof of this fact). The complexity of this model will depend on the complexity of the types, so one way to try to obtain a good effective version of this fact is to restrict ourselves to decidable theories in which each type is computable. A set $X$ is *saturated bounding* if every complete decidable theory each of whose types is computable has an $X$-decidable saturated model.

As in the case of atomic models, we need to draw the distinction between the hypothesis that each type of a complete theory $T$ is computable and the hypothesis that the types of $T$ can be computably listed. In the latter case, we have the following theorem.

**Theorem 9.73** (Morley [149], Millar [139,140]). *A complete decidable theory $T$ has a decidable saturated model iff there is a computable listing of the types of $T$.*

There is a connection between being saturated bounding and being able to subuniformly enumerate the computable sets, in the sense of Theorem 3.23, but we do not know how close this connection is.

$\boxed{80}$ **Exercise 9.74** (Harris [80,81]). Use Theorems 3.23 and 9.73 to show that if $X$ is either high or has PA degree, then $X$ is saturated bounding.

▶ **Open Question 9.75.** Is it the case that $X$ is saturated bounding iff $X$ is either high or has PA degree?

Some evidence for a positive answer to this question is provided by the result of Montalbán [unpublished], building on work of Harris [80,81], that the saturated bounding c.e. sets are exactly the high ones. (A proof of Montalbán's result appears in a later version of [81], which at the time of writing does not appear to be publicly available.)

# Chapter 10

# Charging Ahead: Further Topics

In this final chapter, we discuss a small selection of topics in the computability theoretic and reverse mathematical analysis of combinatorial principles. Included are examples that illustrate some important issues and ideas in the area that were not emphasized in previous chapters, as well as some open questions of current research interest.

## 10.1 The Dushnik-Miller Theorem

A *nontrivial self-embedding* of a linear order $L$ is an order preserving map from $L$ into itself that is not the identity. The *Dushnik-Miller Theorem* states that every infinite linear order has a nontrivial self-embedding. In the countable case, this theorem is easy to prove: If the infinite linear order $L$ has an interval $I$ of type $\omega$ or $\omega^*$, then we can easily define a nontrivial self-embedding of $I$ and extend that to a nontrivial self-embedding of $L$ by defining it to be the identity outside of $I$. Otherwise, it is not difficult to show that $L$ embeds a copy of the rationals. By taking a dense subset, we may assume this copy is a proper suborder $Q$ of $L$. We can then embed $L$ into $Q$, which yields a nontrivial self-embedding of $L$. Nevertheless, the Dushnik-Miller Theorem is not effectively true, and hence is not provable in $RCA_0$, even for the simple case of order type $\omega$:

81 **Exercise 10.1** (Hay and Rosenstein, see [172]). Show that there is a computable linear order $\preccurlyeq$ of type $\omega$ with no computable nontrivial self-embedding. [Hint: To ensure that $\Phi_e$ is not a nontrivial self-embedding of $\preccurlyeq$, we can take an $x$ such that $x \prec \Phi_e(x) \prec \Phi_e(\Phi_e(x))$ and ensure that there are more elements between $x$ and $\Phi_e(x)$ than between $\Phi_e(x)$ and $\Phi_e(\Phi_e(x))$.]

Indeed, Downey and Lempp [44] built a computable linear order $L$ of type $\omega$ such that any nontrivial self-embedding of $L$ computes $\emptyset'$. Their construction can be transformed into a proof that the Dushnik-Miller Theorem implies $\mathrm{ACA}_0$ over $\mathrm{RCA}_0$ (see Downey, Jockusch, and Miller [43]), and hence is equivalent to $\mathrm{ACA}_0$, by Exercise 10.2 below. For order type $\omega$, the result of Downey and Lempp is the best possible, since given a computable linear order $\preccurlyeq$ of type $\omega$, we can use $\emptyset'$ to compute the sequence $a_0 \prec a_1 \prec \cdots$ of elements of $\preccurlyeq$ in order, from which we get the $\emptyset'$-computable nontrivial self-embedding $a_n \mapsto a_{n+1}$. More generally, we have the following fact:

$\boxed{82}$ **Exercise 10.2** (Downey, Jockusch, and Miller [43]). Let $L$ be a computable infinite linear order. Let $A$ be the adjacency relation of $L$, that is, the set of pairs $x \neq y$ in $L$ such that there is no element of $L$ strictly between $x$ and $y$. Let $B$ the block relation of $L$, that is, the set of pairs $x, y$ in $L$ such that there are only finitely many elements of $L$ between $x$ and $y$. Show that $L$ has an $(A \oplus B)$-computable nontrivial self-embedding. Conclude that every computable infinite linear order has an $\emptyset''$-computable nontrivial self-embedding, and that the Dushnik-Miller Theorem is provable in $\mathrm{ACA}_0$.

Downey, Jockusch, and Miller [43] showed that there is a computable infinite linear order with no $\emptyset'$-computable nontrivial self-embedding.

▶ **Open Question 10.3.** Is there a computable infinite linear order $L$ such that every nontrivial self-embedding of $L$ computes $\emptyset''$?

In computable mathematics, we are often interested in properties that are "intrinsic" to a structure, as opposed to a particular computable copy of the structure. (A computable copy of a structure $\mathcal{M}$, that is, a computable structure that is isomorphic to $\mathcal{M}$, is often called a *computable presentation* of $\mathcal{M}$.) For instance, while there are computable copies of $\omega$ with computable nontrivial self-embeddings, we have seen that not all computable copies of $\omega$ have this property. On the other hand, any computable copy of $\mathbb{Q}$ has a computable nontrivial self-embedding (indeed any copy $L$ of $\mathbb{Q}$ has a nontrivial self-embedding computable from $L$), so we can say that the property of having computable nontrivial self-embeddings is intrinsic to $\mathbb{Q}$ but not to $\omega$.

A linear order $L$ is *strongly $\eta$-like* if there is an $n$ such that the order type of $L$ can be obtained from $\eta$ (the order type of the rationals) by replacing each point by a nonempty block of $\leqslant n$ many points. According to Downey [39], the following fact was probably first noted by Watnick and Lerman.

$\boxed{83}$ **Exercise 10.4.** Show that if a computable linear order has a strongly $\eta$-like interval, then it has a computable nontrivial self-embedding.

Thus, if a linear order $L$ has a strongly $\eta$-like interval, then every computable copy of $L$ has a computable nontrivial self-embedding. The long-standing *Effective Dushnik-Miller Conjecture* is that the following question has a positive answer.

▶ **Open Question 10.5.** If every computable copy of a computable linear order $L$ has a computable nontrivial self-embedding, must $L$ contain a strongly $\eta$-like interval?

See [39] for a discussion of the difficulties involved in proving this conjecture.

## 10.2  Linearizing well-founded partial orders

A *linear extension* of a partial order $\preccurlyeq_P$ is a linear order $\preccurlyeq_L$ with the same domain such that if $x \preccurlyeq_P y$ then $x \preccurlyeq_L y$. Part of Exercise 9.6 was to prove the effective version of Szpilrajn's Theorem, which states that every partial order has a linear extension. A simple proof in the countable case (which also establishes the effective version) is as follows. We can assume the domain of $\preccurlyeq_P$ is $\mathbb{N}$ (the finite case is similar). We build $\preccurlyeq_L$ by recursion. Suppose that we have defined $\preccurlyeq_L$ on numbers $< n$ to be a linear extension of $\preccurlyeq_P$ on numbers $< n$. The idea is to place $n$ in $\preccurlyeq_L$ as far to the right as possible.

In greater detail, we proceed as follows: If there is no $m < n$ such that $n \prec_P m$, then let $k \prec_L n$ for all $k < n$, which clearly ensures that $\preccurlyeq_L$ on numbers $\leqslant n$ is a linear extension of $\preccurlyeq_P$ on numbers $\leqslant n$. Otherwise, let $m$ be the $\preccurlyeq_L$-least number such that $n \prec_P m$, and place $n$ immediately before $m$; that is, let $n \prec_L k$ for all $k < n$ such that $m \preccurlyeq_L k$, and let $k \prec_L n$ for all other $k < n$. If $k < n$ and $k \prec_p n$, then $k \prec_p m$, so $k \prec_L m$, and hence $k \prec_L n$. If $k < n$ and $n \prec_p k$ then $m \preccurlyeq_L k$, by the minimality of $m$, so $n \prec_L k$. Thus, in this case too, $\preccurlyeq_L$ on numbers $\leqslant n$ is a linear extension of $\preccurlyeq_P$ on numbers $\leqslant n$. So, in the end, $\preccurlyeq_L$ is a linear extension of $\preccurlyeq_P$.

If $\preccurlyeq_P$ is computable, then so is $\preccurlyeq_L$, and indeed, the above proof can clearly be carried out in $\mathrm{RCA}_0$. But what if we want to preserve some of the structure of $\preccurlyeq_P$? An example of a property of a partial order that could be quite useful to have preserved in a linearization is well-foundedness. As it turns out, the above construction already preserves well-foundedness.

**Theorem 10.6** (Bonnet [6]). *Every well-founded countable partial order can be extended to a well-order.*

*Proof.* Let $\preccurlyeq_P$ be a partial order with domain $\mathbb{N}$ and let $\preccurlyeq_L$ be built as above. Suppose that there is an infinite descending sequence $a_0 \succ_L a_1 \succ_L \cdots$. We use this sequence to obtain an infinite descending sequence $b_0 \succ_P b_1 \succ_P \cdots$. By passing to a subsequence if necessary, we may assume that $a_0 < a_1 < \cdots$. For each $i < j$, either $a_i \succ_P a_j$ or $a_i \mid_P a_j$ (where the latter notation means that $a_i$ and $a_j$ are $\preccurlyeq_P$-incomparable). By Ramsey's Theorem, there is either an infinite subsequence of $\preccurlyeq_P$-comparable elements, or an infinite subsequence of $\preccurlyeq_P$-incomparable ones. In the former case, we have our infinite descending $\preccurlyeq_P$-sequence, as desired. Thus we may assume that there is an infinite subsequence of $\preccurlyeq_P$-incomparable elements. Passing to this subsequence, we may assume that $a_i \mid_P a_j$ for all $i \neq j$.

Let $n > 0$. Since $a_n \prec_L a_0$, by the definition of $\preccurlyeq_L$ there must be a $c < a_n$ such that $a_n \prec_P c \prec_L a_0$. If $c > a_0$, then we can repeat the argument with $c$ in place of $a_n$ (since we cannot have $c \prec_P a_0$). Iterating this argument eventually yields a $c_n$ such that $a_n \prec_P c_n \prec_L a_0$ and $c_n < a_0$. There must be some $c$ such that $c_n = c$ for infinitely many $n$. Let $b_0$ be the least $c < a_0$ (with respect to the usual ordering $<$ on $\mathbb{N}$) such that $a_n \prec_P c \prec_L a_0$ for infinitely many $n$. Let $I_1$ be the set of all $n$ such that $a_n \prec_P b_0$, and let $i_1 = \min I_1$.

Now we can repeat the above argument to obtain a $b_1$ that is the $<$-least $c < a_{i_1}$ such that $a_n \prec_P c \prec_L a_{i_1}$ for infinitely many $n \in I_1$. Let $I_2$ be the set of all $n \in I_1$ such that $a_n \prec_P b_1$, and let $i_2 = \min I_2$. Since $a_{i_1} \prec_P b_0$, we must have $b_1 \neq b_0$. In fact, by the $<$-minimality of $b_0$, we have $b_1 > b_0$. We now keep iterating this process to obtain $b_0 < b_1 < \cdots$. We claim that $b_0 \succ_P b_1 \succ_P \cdots$, as desired.

To verify this claim, suppose that $b_k \not\succ_P b_{k+1}$. We have $b_{k+1} \prec_L a_{i_{k+1}} \prec_P a_{i_k}$, so $b_{k+1} \prec_L b_k$, and hence $b_{k+1} \mid_P b_k$. Since $b_{k+1} > b_k$, we can argue as above to obtain a $d < b_k$ such that $b_{k+1} \prec_P d \prec_L b_k$. But this fact contradicts the $<$-minimality of $b_{k+1}$. (That is, in defining our sequence of $b$'s, we should have picked $b_{k+1}$ to be $\leqslant d$.) $\qquad\square$

Since the linearization $\preccurlyeq_L$ in the above argument is obtained effectively from $\preccurlyeq_P$, we have the following result.

**Theorem 10.7** (Kierstead and Rosenstein, see [173]). *Every well-founded computable partial order has a well-ordered computable linear extension.*

We denote the statement that every well-founded partial order has a well-ordered linear extension by $\mathrm{EXT}(\omega^*)$. (This statement is part of an interesting family of combinatorial statements that we will discuss at the end of this section.) Theorem 10.7 can be seen as an effective version of Theorem 10.6, but should we therefore conclude that $\mathrm{EXT}(\omega^*)$ holds effectively? Here we need to proceed with some caution. There is an important difference between $\mathrm{EXT}(\omega^*)$ and other principles we have been considering. It is true that $\mathrm{EXT}(\omega^*)$ has the usual form $\forall X \, [\theta(X) \to \exists Y \, \varphi(X, Y)]$. However, here $\theta$ and $\varphi$ are not arithmetic. Saying that an order is well-founded requires a set quantifier (indeed, well-foundedness is a $\Pi_1^1$-complete property). Thus we should distinguish between well-founded computable orders and *computably* well-founded computable orders, i.e., computable orders with no computable infinite descending sequences. This distinction does make a difference to the issue of the effectivity of $\mathrm{EXT}(\omega^*)$.

**Theorem 10.8** (Rosenstein and Statman, see [173]). *There is a computably well-founded computable partial order with no computably well-founded computable linear extension.*

We will prove this theorem below, but let us first say a little more about the reverse mathematical import of the form of the statement $\mathrm{EXT}(\omega^*)$. For a true principle $P$ expressed by a sentence of the form $\forall X \, [\theta(X) \to \exists Y \, \varphi(X, Y)]$, a proof that

$$\forall \text{ computable } X \, [\theta(X) \to \exists \text{ computable } Y \, \varphi(X, Y)] \qquad (10.1)$$

can often be transformed into a proof of $P$ in $\mathrm{RCA}_0$, or at least in $\mathrm{RCA}_0$ together with some additional amount of induction. Of course, that is not always the case. The proof that $\varphi(X, Y)$ holds may need additional power, or the result may not be relativizable. Even in such cases, though, as pointed out in Section 4.5, if $\theta$ and $\varphi$ are arithmetic, then we may conclude that $P$ holds in the minimal $\omega$-model $\mathcal{M}$ of $\mathrm{RCA}_0$ (which consists of the computable sets), and hence cannot imply any principle that is not true in that model (which includes every principle stronger than $\mathrm{RCA}_0$ that we have discussed, except for bounding and induction principles and their consequences). Without the assumption of arithmeticity, though, this need not be the case. For instance, there may be computable $X$ for which

$\mathcal{M} \vDash \theta(X)$ but $\theta(X)$ does not in fact hold, so that (10.1) has nothing to say about whether there is a computable $Y$ such that $\varphi(X, Y)$ holds (or holds in $\mathcal{M}$). Theorems 10.7 and 10.8 show that $\text{EXT}(\omega^*)$ is an example of this situation (we will see another in Section 10.3). For this principle, (10.1) holds, but the principle is not true in $\mathcal{M}$.

Thus $\text{EXT}(\omega^*)$ does not hold in $\text{RCA}_0$. In fact, we have the following result, which can be proved simultaneously with Theorem 10.8.

**Theorem 10.9** (Downey, Hirschfeldt, Lempp, and Solomon [42]). $\text{EXT}(\omega^*)$ *implies* $\text{WKL}_0$ *over* $\text{RCA}_0$.

*Proof of Theorems 10.8 and 10.9.* Let $T$ be an infinite binary tree. For $\sigma, \tau \in T$, let $\sigma \preccurlyeq_P \tau$ iff $\sigma$ extends $\tau$. We show that any linear extension $\preccurlyeq_L$ of $\preccurlyeq_P$ has an infinite descending sequence.

We proceed by recursion. Let $c_0$ be the empty string. Suppose we have defined $c_0 \succ_L c_1 \succ_L \cdots \succ_L c_n$. Let $C_n = \{c_i : i \leqslant n\}$ and let $D_n$ be the set of immediate successors (in $T$) of elements of $C_n$. Since $T$ is infinite and $c_0 \in C_n$, we cannot have $D_n \subseteq C_n$, since then we could show by induction on the length of $\sigma$ that every $\sigma \in T$ is in $C_n$. Let $c_{n+1}$ be the $\preccurlyeq_L$-greatest element of $D_n \setminus C_n$. Then $c_{n+1} \prec_L c_n$, since otherwise $c_{n+1}$ is not an immediate successor of $c_n$, and hence $c_{n+1} \in D_{n-1}$, so $c_{n+1}$ should have been picked to be $c_n$ (or $n = 0$, in which case $c_{n+1} \prec_L c_n$ is automatic).

The above argument works in $\text{RCA}_0$. If we also have $\text{EXT}(\omega^*)$ then we may conclude that $\preccurlyeq_P$ also has an infinite descending sequence, from which we obtain a path on $T$, thus establishing Theorem 10.9.

Furthermore, the sequence $c_0, c_1, \ldots$ is built computably from $\preccurlyeq_L$, so if we take $T$ to be a computable infinite binary tree with no computable path, then Theorem 10.8 follows.                                                    $\square$

On the other hand, we do have the following fact.

$\boxed{84}$ **Exercise 10.10** (Downey, Hirschfeldt, Lempp, and Solomon [42]). Verify that the argument in Theorem 10.6 can be carried out in $\text{ACA}_0$, and hence $\text{ACA}_0 \vdash \text{EXT}(\omega^*)$.

▶ **Open Question 10.11.** Does $\text{EXT}(\omega^*)$ imply $\text{ACA}_0$ over $\text{RCA}_0$? What is the relationship between $\text{EXT}(\omega^*)$ and $\text{RT}_2^2$?

As mentioned above, $\text{EXT}(\omega^*)$ is part of a family of combinatorial statements. We use the same notation for order types as in Section 9.1. We say that an order type $\tau$ is *strongly extendible* if every partial order with no linear suborder of type $\tau$ has a linear extension with no suborder of type $\tau$. In this definition, the universal quantifier ranges over all partial orders, including uncountable ones, so this notion is not amenable to treatment in the framework of classical computability theory and reverse mathematics. Thus we work with the following notion, studied by Jullien [108], among others.

**Definition 10.12.** Let $\tau$ be an order type. We say that $\tau$ is *weakly extendible* if every countable partial order with no linear suborder of type $\tau$ has a linear extension with no suborder of type $\tau$. Let $\mathrm{EXT}(\tau)$ be the statement that $\tau$ is weakly extendible.

Not every order type is weakly extendible. The countably infinite partial order in which any two elements are incomparable witnesses the fact that if $\tau$ is weakly extendible and has more than one element, then it must be infinite. More generally, for any countable order type $\rho$, the partial order consisting of $\rho$ followed by infinitely many incomparable elements witnesses the failure of $\mathrm{EXT}(\rho + n)$ for any $n > 1$. A different sort of example is $\mathrm{EXT}(\omega + \omega^*)$, which also fails to hold. There are also countable order types that are weakly extendible but not strongly extendible, such as $\omega + 1$. For these and other classical facts on extendibility, see Bonnet and Pouzet [7]. Note also that $\mathrm{EXT}(\tau)$ is clearly equivalent to $\mathrm{EXT}(\tau^*)$.

There are characterizations of the strongly extendible and weakly extendible order types, due to Bonnet [6] and Jullien [108], respectively. We will briefly discuss Jullien's Theorem in Section 10.3. In addition to $\mathrm{EXT}(\omega^*)$, other special cases have been studied from the computability theoretic and reverse mathematical points of view, including $\mathrm{EXT}(\zeta)$ and $\mathrm{EXT}(\eta)$, both of which we will discuss in Section 10.3. (Here $\zeta$ is the order type of the integers and $\eta$ the order type of the rationals.)

## 10.3  The world above $\mathrm{ACA}_0$

In this section, we look at a small sampling of the many interesting principles that live above $\mathrm{ACA}_0$ in the reverse mathematical universe.

### 10.3.1  $\mathrm{ATR}_0$ and $\Pi_1^1\text{-}\mathrm{CA}_0$

In addition to $\mathrm{RCA}_0$, $\mathrm{WKL}_0$, and $\mathrm{ACA}_0$, there are two other systems that make up what are usually considered the "big five" systems of reverse mathematics. One is $\Pi_1^1$-$CA_0$, which consists of $RCA_0$ together with the $\Pi_1^1$-comprehension scheme (see Definition 4.2). The other is $ATR_0$, which consists of $ACA_0$ together with an axiom scheme stating that arithmetic comprehension can be iterated along any countable well order. In greater detail, we may state this axiom scheme as follows: Let $\varphi(n, X)$ be an arithmetic formula. Here $n$ and $X$ are free variables of $\varphi$, but $\varphi$ could contain additional free variables (i.e., parameters). Let $\Phi(X) = \{n : \varphi(n, X)\}$. Let $\preceq$ be a well order with domain $D$. Then there is a sequence of sets $\{K_n : n \in D\}$ such that

$$K_n = \Phi(\{\langle m, x \rangle : m \prec n \wedge x \in K_m\}).$$

Equivalents of $ATR_0$ include the fact that any two countable well-orders are comparable, Ulm's Theorem on Abelian $p$-groups, and Lusin's Separation Theorem. Equivalents of $\Pi_1^1$-$CA_0$ include the Cantor-Bendixson Theorem and Silver's Theorem. See Simpson [190, 191] for the precise statements of these theorems and longer lists of equivalents of $ATR_0$ and $\Pi_1^1$-$CA_0$, and [191] for proofs of these equivalences and the relevant references.

$\boxed{85}$ **Exercise 10.13** (Friedman [59]). Show that $ACA_0 \nvdash ATR_0$. [Hint: Recall that the arithmetic sets form an $\omega$-model of $ACA_0$ and use $ATR_0$ to construct a nonarithmetic set.]

By definition, $ATR_0$ implies $ACA_0$. The following fact completes the proof that the big five systems of reverse mathematics are linearly ordered by strength.

$\boxed{86}$ **Exercise 10.14** (Friedman [59]). Show that $\Pi_1^1$-$CA_0 \vdash ATR_0$.

See [191] for a proof that $ATR_0 \nvdash \Pi_1^1$-$CA_0$.

One may think of $ATR_0$ as a higher-level analog to $WKL_0$. For example, as in the case of $WKL_0$, there is no minimal $\omega$-model of $ATR_0$, but the intersection of all $\omega$-models of $ATR_0$ consists exactly of the hyperarithmetic (i.e., $\Delta_1^1$) sets, in the same way that the intersection of all $\omega$-models of $WKL_0$ consists of the computable (i.e., $\Delta_1^0$) sets (see [191]). Similarly, recall from Exercise 4.16 that $WKL_0$ is equivalent to the $\Sigma_1^0$-separation principle; the following theorem states that $ATR_0$ is equivalent to the $\Sigma_1^1$-separation principle (for a proof, see Theorem V.5.1 of [191]).

**Theorem 10.15** (see Simpson [191]). *$ATR_0$ is equivalent over $RCA_0$ to the $\Sigma_1^1$-separation principle, which states that for any $\Sigma_1^1$ formulas $\varphi(n)$ and $\psi(n)$ in which $X$ does not occur free, if $\neg \exists n [\varphi(n) \wedge \psi(n)]$ then*

$$\exists X [(\varphi(n) \to n \in X) \wedge (\psi(n) \to n \notin X)].$$

In this analogy, $\Pi_1^1$-CA$_0$ plays the role of ACA$_0$, while the role of RCA$_0$ is taken by the less frequently encountered system $\Delta_1^1$-CA$_0$. See [191] for more details.

We will not discuss $\Pi_1^1$-CA$_0$ much here (Simpson [191] deals with it extensively), but the following remark is worth keeping in mind when working with principles around the level of $\Pi_1^1$-CA$_0$. A $\beta$-*model* is an $\omega$-model $\mathcal{M}$ such that, for every $\Sigma_1^1$ formula $\varphi$ with parameters from $\mathcal{M}$, we have $\mathcal{M} \vDash \varphi$ iff $\varphi$ is true (i.e., $\varphi$ holds in the standard model $(\omega, \mathcal{P}(\omega), +, \cdot, 0, 1, \leqslant)$ of second order arithmetic). Chapter VII of [191] is devoted to $\beta$-models; in Section 1 of that chapter, it is shown that $\Pi_1^1$-CA$_0$ has a minimum $\beta$-model, but ATR$_0$ does not. In particular, there are $\beta$-models of ATR$_0$ that are not models of $\Pi_1^1$-CA$_0$. On the other hand, it is easy to see that any true $\Pi_2^1$ statement (which includes almost all of the principles discussed in this book) holds in every $\beta$-model. Therefore we have the following fact.

**Proposition 10.16** (see Simpson [191]). *No true $\Pi_2^1$ statement can imply $\Pi_1^1$-CA$_0$, even over ATR$_0$.*

There is also work on higher levels of the reverse mathematical universe (e.g., $\Pi_2^1$-comprehension), some of which is described in [191]. Recent examples include [51, 148, 152].

## 10.3.2 The extendibility of $\zeta$ and $\eta$

Recall the notion of weak extendibility from Definition 10.12. As we will not discuss strong extendibility further, we will henceforth use "extendible" to mean weakly extendible. The extendibility of $\zeta$ yields a relatively straightforward example of a proof of equivalence to ATR$_0$. The proof of ATR$_0$ from EXT($\zeta$) uses a two-step process that is common in reversals to ATR$_0$, together with the following useful lemma, a proof of which can be found in Section V.1 of [191].

**Lemma 10.17.** *The following is provable in ACA$_0$. Let $\varphi(x)$ be a $\Pi_1^1$ formula. Then there is a sequence of trees $T_0, T_1, \ldots$ such that $\varphi(n)$ iff $T_n$ has no path.*

**Theorem 10.18** (Downey, Hirschfeldt, Lempp, and Solomon [41]). EXT($\zeta$) *is equivalent to* ATR$_0$ *over* RCA$_0$.

*Proof.* We first show that Jullien's proof in [108] that $\zeta$ is extendible can be carried out in ATR$_0$: Let $P$ be a partial order that avoids $\zeta$. We may assume that the domain of $P$ is $\mathbb{N}$. Say that $a$ is $<$-*good* if $P$ restricted to the predecessors of $a$ avoids $\omega^*$, and that $a$ is $>$-*good* if $P$ restricted to the successors of $a$ avoids $\omega$. Every element of $P$ is either $<$-good or $>$-good (or both). Both $<$-goodness and $>$-goodness are $\Pi^1_1$ properties, so by Theorem 10.15, there is a set $S$ such that every element of $S$ is $<$-good, and every element of $\overline{S}$ is $>$-good. Since the set of $<$-good numbers is closed downward in $P$, we may assume that so is $S$. The restriction $P_0$ of $P$ to $S$ avoids $\omega^*$, and the restriction $P_1$ of $P$ to $\overline{S}$ avoids $\omega$, so by Exercise 10.10, $P_0$ has a linearization $L_0$ that avoids $\omega^*$, and $P_1$ has a linearization $L_1$ that avoids $\omega$. Putting all elements of $L_0$ before all elements of $L_1$ yields a linearization of $P$ that avoids $\zeta$.

For the reversal, we proceed in two steps. First we use the extendibility of $\zeta$ to prove $\mathrm{ACA}_0$ over $\mathrm{RCA}_0$. Then we use it again to prove $\mathrm{ATR}_0$ over $\mathrm{ACA}_0$. So henceforth we assume the extendibility of $\zeta$.

We first argue in $\mathrm{RCA}_0$. Let $f : \mathbb{N} \to \mathbb{N}$. We show that the range of $f$ exists. Build a partial order $P$ as follows. (We give an informal description, but it is not difficult to formalize it.) For each $n$, begin with two elements $a_n$ and $b_n$, and declare all the $a_n$'s and $b_n$'s to be $P$-incomparable. For each $n$, begin to build a chain of order type $\omega$ above $a_n$ and a chain of order type $\omega^*$ below $b_n$. If we ever find an $m$ such that $f(n) = m$, then switch to building a chain of order type $\omega^*$ below $a_n$ and a chain of order type $\omega$ above $b_n$ (which of course leaves finitely many elements above $a_n$ and finitely many elements below $b_n$).

Clearly, $P$ avoids $\zeta$, so it has a linearization $\preccurlyeq$ that avoids $\zeta$. If $n \notin \mathrm{rng}\, f$, then $a_n \prec b_n$, as otherwise the $\omega^*$-chain below $b_n$ would precede the $\omega$-chain above $a_n$. Similarly, if $n \in \mathrm{rng}\, f$ then $b_n \prec a_n$. Thus $\{n : b_n \prec a_n\} = \mathrm{rng}\, f$.

We now argue in $\mathrm{ACA}_0$. Let $\varphi(x)$ and $\psi(x)$ be $\Sigma^1_1$ formulas such that $\forall x\, [\neg(\varphi(x) \wedge \psi(x))]$. By Lemma 10.17, there are sequences of trees $T^0_0, T^0_1, \ldots$ and $T^1_0, T^1_1, \ldots$ such that $\varphi(n)$ iff $T^0_n$ is not well-founded and $\psi(n)$ iff $T^1_n$ is not well-founded. (Invoking this lemma is the reason we need $\mathrm{ACA}_0$ here.) By $(T^i_n)^*$ we mean an upside-down copy of $T^i_n$. Now let $P$ be the partial ordered defined as follows. For each $n$, begin with two elements $a_n$ and $b_n$, and declare all the $a_n$'s and $b_n$'s to be $P$-incomparable. For each $n$, place a copy of $T^0_n$ above $a_n$, a copy of $(T^1_n)^*$ below $a_n$, a copy of $T^1_n$ above $b_n$, and a copy of $(T^0_n)^*$ below $b_n$.

Since for each $n$, at least one of $T^0_n$ and $T^1_n$ is well-founded, $P$ avoids $\zeta$, so it has a linearization $\preccurlyeq$ that avoids $\zeta$. If $\varphi(n)$ holds, then $a_n \prec b_n$, as otherwise the path on $(T^0_n)^*$ below $b_n$ would precede the path on $T^0_n$ above $a_n$. Similarly, if $\psi(n)$ holds then $b_n \prec a_n$. Thus $\{n : a_n \prec b_n\}$ is a set separating $\varphi$ and $\psi$, so by Theorem 10.15, $\text{ATR}_0$ holds. □

As part of his work on Jullien's Theorem discussed below, Montalbán [144] studied the extendibility of particular order types, including $\eta$. (Another way to state $\text{EXT}(\eta)$ is that every scattered partial order has a scattered linear extension.) Becker (see [42]) showed that $\Pi_1^1\text{-CA}_0 \vdash \text{EXT}(\eta)$. Montalbán [144] improved this result by showing that $\text{ATR}_0 + \text{I}\Sigma_1^1 \vdash \text{EXT}(\eta)$. Conversely, Miller [unpublished] showed that $\text{EXT}(\eta)$ implies $\text{WKL}_0$ over a system stronger than $\text{RCA}_0$. The exact strength of $\text{EXT}(\eta)$ is still open. In particular, we have the following questions.

▶ **Open Question 10.19.** Does $\text{ATR}_0 \vdash \text{EXT}(\eta)$? Does $\text{RCA}_0 + \text{EXT}(\eta) \vdash \text{ATR}_0$?

### 10.3.3 Maximal linear extensions

Another example of a principle equivalent to $\text{ATR}_0$ is the maximal linear extension theorem of de Jongh and Parikh [35]. A partial order $(P, \preccurlyeq)$ is a *well-partial-order* (*wpo*) if for every $f : \mathbb{N} \to P$ there are $m < n$ such that $f(m) \preccurlyeq f(n)$ (in other words, there are no infinite descending sequences and no infinite antichains; see Cholak, Marcone, and Solomon [21] for an analysis of the reverse mathematical aspects of equivalent definitions of wpo). A partial order is a wpo iff all its linearizations are well-orders. (See [21] for a proof and reverse mathematical analysis of this fact.) The *maximal order type* $o(P)$ of a wpo $P$ is the supremum of the order types of its linearizations (where we think of these order types as ordinals). The *maximal linear extension theorem* states that every wpo $P$ has a linearization whose order type is $o(P)$. We call such a linearization a *maximal linear extension*. The original proof of this theorem can be carried out in $\text{ATR}_0$.

From a computability theoretic point of view, we have the following result.

**Theorem 10.20** (Montalbán [145]). *Every computable wpo has a computable maximal linear extension.*

However, Montalbán [145] also showed that the problem of finding maximal linear extensions for computable wpo's is not at all easy. Specifically, he showed that there is no hyperarithmetic way to compute an index for a computable maximal linear extension of a computable wpo $P$ given an index for $P$. More recently, the reverse mathematical strength of the maximal linear extension theorem was precisely determined.

**Theorem 10.21** (Marcone and Shore [131]). *The maximal linear extension theorem is equivalent to* $\mathrm{ATR}_0$ *over* $\mathrm{RCA}_0$.

In particular, the maximal linear extension theorem does not hold in the minimal $\omega$-model $\mathcal{M}$ of $\mathrm{RCA}_0$. The issue here is similar to what we encountered in Section 10.2: there are computable partial orders $(P, \preccurlyeq)$ that "look like" wpo's from the point of view of $\mathcal{M}$ (i.e., for every *computable* $f : \mathbb{N} \to P$ there are $m < n$ such that $f(m) \preccurlyeq f(n)$), but are not in fact wpo's.

### 10.3.4  Kruskal's Theorem, Fraïssé's Conjecture, and Jullien's Theorem

A *quasiorder* on a set $P$ is a binary relation on $P$ that is reflexive and transitive. A quasiorder $(P, \preccurlyeq)$ is a *well-quasiorder* (*wqo*) if for every $f : \mathbb{N} \to P$ there are $m < n$ such that $f(m) \preccurlyeq f(n)$ (so the antisymmetric wqo's are exactly the wpo's). Marcone [129] is a survey of the reverse mathematics of the theory of wqo's, and of the considerably more complicated notion of a bqo (better quasiorder); see also Montalbán [146]. (The complexity of the latter notion is demonstrated by the fact that the only known proof that the partial ordering consisting of three incomparable elements is a bqo requires $\mathrm{ATR}_0$, and it is not known whether this statement can be proven in any system weaker than $\mathrm{ATR}_0$.) Two well known results involving wqo's are Kruskal's Theorem [117] that the finite trees form a wqo under the embeddability relation (preserving greatest lower bounds) and Laver's Theorem [123] that the countable linear orders form a wqo under the embeddability relation.

The usual proof of Kruskal's Theorem proceeds by assuming that there is an infinite sequence of trees violating the definition of wqo, and then using the existence of a minimal such sequence. Such an argument needs some form of $\Pi_1^1$-comprehension. Not only is finding different, metamathematically simpler proofs of this result of foundational interest but, as Rathjen and Weiermann [168] put it, the desirability of having such proofs "is especially felt due to the fact that this theorem figures prominently in computer science, because it is the main tool for showing that sets of rewrite rules are terminating [. . . ]" (They quote Gallier [68] as a place "where this challenge is offered".) Friedman (see [186]) showed that Kruskal's Theorem is not provable in $\mathrm{ATR}_0$. Rathjen and Weiermann [168] pinpointed the reverse mathematical strength of Kruskal's Theorem.

Laver's Theorem is still often referred to as Fraïssé's Conjecture. Thus, in the context of reverse mathematics, it is usually denoted by FRA. (Laver actually proved a stronger statement, which is the one that has been called Laver's Theorem and denoted by LAV in the context of reverse mathematics; see [146].) Laver's proof (which uses bqo's) can be carried out in $\Pi_2^1$-CA$_0$ (i.e., RCA$_0$ together with the $\Pi_2^1$-comprehension scheme), but by Proposition 10.16, FRA cannot imply $\Pi_1^1$-CA$_0$. Here again we have the challenge of finding reverse mathematically simpler proofs of an important theorem whose original proof uses quite complex methods. Shore [182] showed that FRA implies ATR$_0$ over RCA$_0$, but the following question remains open.

▶ **Open Question 10.22.** Does ATR$_0 \vdash$ FRA?

Several people have conjectured that the answer to this question is positive, but we do not even know whether $\Pi_1^1$-CA$_0 \vdash$ FRA. See Marcone and Montalbán [130] for a recent approach to this question (also summarized in [146]).

Montalbán [144] gave examples of interesting statements equivalent to FRA. One of these is Jullien's Theorem, mentioned above, which classifies the extendible countable order types (although in this case the equivalence requires additional induction). See [144] for a statement of Jullien's Theorem, and the version known as JUL.

**Theorem 10.23** (Montalbán [144]). RCA$_0$ + JUL $\vdash$ FRA *and* RCA$_0$ + I$\Sigma_1^1$ + FRA $\vdash$ JUL.

Montalbán [147] draws the conclusion from his work that FRA "has a *robustness property* in the sense that it is equivalent to many other statements talking about the same type of objects. So far, the only systems with this robustness property were the main five (and WWKL$_0$)..." In [144], he additionally writes that "Simpson claimed [in [191]] that, over RCA$_0$, Friedman's system, ATR$_0$, is the weakest set of axioms which permits the development of a decent theory of countable ordinals. Similarly, we should conclude from our work that, over [RCA$_0$ + I$\Sigma_1^1$], FRA [...] is the weakest set of axioms which permits the development of a decent theory of countable linear orderings modulo equimorphisms." Thus a negative solution to Question 10.22 could be of at least as much interest as a positive one, revealing a new robust subsystem of second order arithmetic, and one with a clear combinatorial motivation and role.

There is a great deal more work that has been done in this fascinating part of the reverse mathematical universe. The papers cited above are a good starting point for further study; other related papers include those by Friedman, Robertson, and Seymour [65] on the celebrated Robertson-Seymour graph minor theorem, and by Neeman [154, 155] on Jullien's indecomposability theorem. A particularly interesting aspect of [155] is that it gives an example of a reverse mathematical implication that provably requires a high level of induction.

### 10.3.5  Hindman's Theorem

There are also principles that may be provable in $ACA_0$, but have not yet been shown to be so. A prominent example is Hindman's Theorem, another of the many Ramsey theoretic principles that have proved to be of reverse mathematical interest. For $X \subseteq \mathbb{N}$, let $S(X)$ be the set of sums of finite nonempty subsets of $X$. *Hindman's Theorem* [83] states that if we partition $\mathbb{N}$ into sets $C_0, \ldots, C_n$, then there is an infinite $X \subseteq \mathbb{N}$ such that $S(X) \subseteq C_i$ for some $i \leqslant n$. Blass, Hirst, and Simpson [5] showed that Hindman's Theorem implies $ACA_0$ over $RCA_0$, and can be proved in the system $ACA_0^+$ introduced at end of Section 6.3, which is strictly stronger than $ACA_0$ but weaker than $ATR_0$.

▶ **Open Question 10.24.** Is Hindman's Theorem provable in $ACA_0$?

Hindman's Theorem has several proofs. Combinatorial proofs (such as Hindman's original one) were studied by Blass, Hirst, and Simpson [5] and Towsner [202]. Another proof uses ultrafilters, and was examined from the point of view of reverse mathematics by Hirst [91] and Towsner [201]. Yet another proof involves topological dynamics, using a theorem that, like Hindman's Theorem itself, follows from $ACA_0^+$ and may be equivalent to $ACA_0$; see [5, 146] for details. Further computability theoretic questions related to Hindman's Theorem can be found in Blass [4]. Hirst [93] analyzed a statement that seems quite similar to (an equivalent form of) Hindman's Theorem, but is much weaker, being equivalent to $B\Sigma_2^0$ over $RCA_0$. The analysis of different proofs and versions of Hindman's Theorem in these papers makes for an excellent case study in how reverse mathematics can reveal fascinating nuances that help us understand the nature of various combinatorial principles and proof techniques.

## 10.4   Still further topics, and a final exercise

There are several other combinatorial principles that have been studied from the computability theoretic and reverse mathematical viewpoints. The following are a few examples with connections to the concepts studied in this book. This brief list is intended as a jumping off point for further exploration, and is by no means exhaustive. Similarly, the lists of references accompanying each topic are merely pointers to the literature, and also not meant to be exhaustive. (See also Section 1.5.)

Several other versions of Ramsey's Theorem have been considered. We have already mentioned the Canonical Ramsey Theorem [137, 138] and the Polarized Ramsey Theorem [22, 48]; other examples include the Dual Ramsey Theorem [143], the Rainbow Ramsey Theorem [29, 34, 38, 142, 205–207], and "infinite exponent" versions discussed in Sections V.9, VI.6, and VI.7 of Simpson [191]. Gasarch [personal communication] has pointed out that the work of Mileti [137, 138] on the Canonical Ramsey Theorem (CRT) is a particularly good example of how computability theoretic considerations can lead to new and independently interesting proofs of known results. To obtain tight computability theoretic bounds on the complexity of versions of CRT, Mileti gave new proofs of CRT. These have been mined by Gasarch [unpublished] to obtain bounds on the Canonical Ramsey numbers arising from finite versions of CRT that, while not optimal, are much easier to obtain than previously known ones.

Of course, there are other Ramsey theoretic principles still awaiting computability theoretic and reverse mathematical analysis. One interesting example, due to Carlson [13], is mentioned in Montalbán [146]. Miller (see [146]) suggested the reverse mathematical study of a family of principles introduced by Erdős, Hajnal, and Rado [54]. These principles are of the form $RT^n_{k,j}$, stating that for each $k$-coloring $c$ of $[\mathbb{N}]^n$, there is an infinite set $S$ such that the restriction of $c$ to $[S]^n$ uses only $j$ many colors. Montalbán [146] notes that Lempp, Miller, and Ng observed that $RT^3_{3,2}$ implies $RT^2_2$, and Wang [207] has shown that for each $n$ there is a $j$ such that $RCA_0 + RT^n_{k,j} \nvdash ACA_0$ for all $k > j$. Dorais, Dzhafarov, Hirst, Mileti, and Shafer [38] have studied the $RT^n_{k,k-1}$ case, which they think of as a variant of the Thin Set Theorem and denote by $TS^n_k$ ($TS(n)$ being $TS^n_\omega$ in their notation).

Ramsey theoretic principles on structures other than the natural numbers are also of interest. An example is versions of Ramsey's Theorem on trees [27, 30, 49, 134]. Hummel and Jockusch [95] studied generalizations of the notion of cohesiveness. Dzhafarov [45] studied "almost all" versions of

$\mathrm{SRT}_2^2$ using tools from effective measure theory. Flood [56] studied a principle that is a hybrid of Weak König's Lemma and Ramsey's Theorem for pairs. Combinatorial principles of reverse mathematical interest can come from many different sources. Dzhafarov [46] gave an example arising from mathematical psychology.

$\boxed{87}$ **Exercise 10.25.** Find a theorem or (even better) family of theorems not yet studied from the computability theoretic and reverse mathematical points of view, and undertake such a study. If the results are sufficiently interesting, publish them.

# Appendix

# Lagniappe: A Proof of Liu's Theorem

In this appendix, we prove Liu's Theorem 6.71 that $\mathrm{RT}_2^2$ does not imply $\mathrm{WKL}_0$ over $\mathrm{RCA}_0$. This proof includes a simplification of the proof of Lemma A.4 below due to Wang [personal communication]. Let $\mathcal{C}$ be the class of sets that are not of PA degree. By taking $T$ such that $[T]$ is a universal $\Pi_1^0$ class in Theorem 6.50, we see that for each $C \in \mathcal{C}$, each family of $C$-computable sets has a cohesive set $B$ such that $B \oplus C \in \mathcal{C}$. We will show that this fact about COH also holds for $\mathrm{SRT}_2^2$. That is, we will show that for each set $C \in \mathcal{C}$ and each set $A$, there is an infinite subset $B$ of $A$ or its complement such that $B \oplus C \in \mathcal{C}$. It then follows from Exercise 4.28 that there is an $\omega$-model of $\mathrm{RCA}_0 + \mathrm{COH} + \mathrm{SRT}_2^2$ (and hence of $\mathrm{RCA}_0 + \mathrm{RT}_2^2$) consisting entirely of sets in $\mathcal{C}$. This model does not contain any paths on $T$, and hence is not a model of $\mathrm{WKL}_0$.

So let $C$ be a set of non-PA degree and let $A$ be any set. Of course, if $C$ can compute $A$ then we are done, so we may assume that $A \not\leq_T C$. We build $G$ such that both $(G \cap A)$ and $(G \cap \overline{A})$ are infinite, and at least one of $(G \cap A) \oplus C$ and $(G \cap \overline{A}) \oplus C$ does not have PA degree.

**The requirements.** To ensure that $(G \cap A)$ and $(G \cap \overline{A})$ are infinite, we will satisfy requirements

$$Q_m : \exists n_0, n_1 > m \, (n_0 \in G \cap A \wedge n_1 \in G \cap \overline{A}).$$

To ensure that $(G \cap A) \oplus C$ does not have PA degree, we would need to satisfy the requirements

$$R_e^A : \Phi_e^{(G \cap A) \oplus C} \text{ total} \Rightarrow \exists n \, \Phi_e^{(G \cap A) \oplus C}(n) \neq \Phi_n(n)\!\downarrow.$$

(Here $\Phi_0, \Phi_1, \dots$ lists all partial computable $0, 1$-valued functionals.) To ensure that $(G \cap \overline{A}) \oplus C$ does not have PA degree, we would need to satisfy the requirements

$$R_i^{\overline{A}} : \Phi_i^{(G \cap \overline{A}) \oplus C} \text{ total} \Rightarrow \exists n \, \Phi_i^{(G \cap \overline{A}) \oplus C}(n) \neq \Phi_n(n)\!\downarrow.$$

193

As in the proof of Theorem 6.57, we employ the standard computability theoretic trick of satisfying the requirements

$$R_{e,i} : R_e^A \vee R_i^{\overline{A}}.$$

**Notation.** Let $\sigma \in 2^{<\mathbb{N}}$ and let $X$ be either an element of $2^{\mathbb{N}}$ or an element of $2^{<\mathbb{N}}$ of length at least the same as that of $\sigma$. We write $\sigma \preccurlyeq X$ to mean that $\sigma$ is an initial segment of $X$, and $\sigma \prec X$ to mean that $\sigma$ is a proper initial segment of $X$. We write $X/\sigma$ for the object obtained by replacing the first $|\sigma|$ many bits of $X$ by $\sigma$. We denote the empty string by $\lambda$.

For sets $X_0, \ldots, X_n$, let $X_0 \oplus \cdots \oplus X_n$, also written as $\bigoplus_{i \leqslant n} X_i$, be $\{(n+1)m + i : m \in X_i\}$. For strings $\tau_0, \ldots, \tau_n$ of the same length, let $\tau_0 \oplus \cdots \oplus \tau_n$ be the string whose $((n+1)m+i)$th bit is $\tau_i(m)$. Note that $\tau_0 \oplus \cdots \oplus \tau_n \prec X_0 \oplus \cdots \oplus X_n$.

**The conditions.** We will use conditions that are elaborations on Mathias forcing conditions. For convenience of notation, we present our Mathias conditions slightly differently from above. Here a *Mathias condition* is a pair $(\sigma, X)$ with $\sigma \in 2^{<\mathbb{N}}$ and $X \in 2^{\mathbb{N}}$. The Mathias condition $(\tau, Y)$ *extends* the Mathias condition $(\sigma, X)$ if $\sigma \preccurlyeq \tau$ and $Y/\tau \subseteq X/\sigma$. A set $G$ *satisfies* the Mathias condition $(\sigma, X)$ if $\sigma \prec G$ and $G \subseteq X/\sigma$.

To define our conditions, we first make the following preliminary definitions. A *$k$-partition of* $\mathbb{N}$ is a set $X = X_0 \oplus \cdots \oplus X_{k-1}$ such that $\bigcup_{i<k} X_i = \mathbb{N}$. (Note that we do not require the $X_i$ to be pairwise disjoint.) A *$k$-partition class* is a nonempty collection of sets, each of which is a $k$-partition of $\mathbb{N}$. Recall that a $\Pi_1^{0,C}$ class is a class that is $\Pi_1^0$ relative to $C$. We will be interested in $\Pi_1^{0,C}$ $k$-partition classes, that is, $\Pi_1^{0,C}$ classes that are also $k$-partition classes.

**Definition A.1.** 1. A *condition* is a tuple of the form $(k, \sigma_0, \ldots, \sigma_{k-1}, P)$, where $k > 0$, each $\sigma_i \in 2^{<\mathbb{N}}$, and $P$ is a $\Pi_1^{0,C}$ $k$-partition class. We think of each $X_0 \oplus \cdots \oplus X_{k-1} \in P$ as representing $k$ many Mathias conditions $(\sigma_i, X_i)$ for $i < k$.

2. A condition $d = (m, \tau_0, \ldots, \tau_{m-1}, Q)$ *extends* $c = (k, \sigma_0, \ldots, \sigma_{k-1}, P)$ if there is a function $f : m \to k$ with the following property: for each $Y_0 \oplus \cdots \oplus Y_{m-1} \in Q$ there is an $X_0 \oplus \cdots \oplus X_{k-1} \in P$ such that each Mathias condition $(\tau_i, Y_i)$ extends the Mathias condition $(\sigma_{f(i)}, X_{f(i)})$. In this case, we say that $f$ *witnesses* this extension, and that *part $i$ of $d$ refines part $f(i)$ of $c$*. (Whenever we say that a condition extends another, we assume we have fixed a function witnessing this extension.)

3. A set $G$ *satisfies* the condition $(k, \sigma_0, \ldots, \sigma_{k-1}, P)$ if there is an $X_0 \oplus \cdots \oplus X_{k-1} \in P$ such that $G$ satisfies some Mathias condition $(\sigma_i, X_i)$. In this case, we also say that $G$ satisfies this condition *on part $i$*. (Note that if $d$ extends $c$ and $G$ satisfies $d$, then $G$ satisfies $c$.)

4. A condition $(k, \sigma_0, \ldots, \sigma_{k-1}, P)$ *forces $Q_m$ on part $i$* if there exist $n_0, n_1 > m$ with $n_0 \in A$ and $n_1 \notin A$ such that $\sigma_i(n_0) = \sigma_i(n_1) = 1$. Clearly, if $G$ satisfies such a condition on part $i$, then $G$ satisfies requirement $Q_m$. (Note that if $c$ forces $Q_m$ on part $i$, and part $j$ of $d$ refines part $i$ of $c$, then $d$ forces $Q_m$ on part $j$.)

5. A condition *forces $R_{e,i}$ on part $j$* if every $G$ satisfying this condition on part $j$ also satisfies requirement $R_{e,i}$. A condition *forces $R_{e,i}$* if it forces $R_{e,i}$ on each of its parts. (Note that if $c$ forces $R_{e,i}$ on part $i$, and part $j$ of $d$ refines part $i$ of $c$, then $d$ forces $R_{e,i}$ on part $j$. Therefore, if $c$ forces $R_{e,i}$ and $d$ extends $c$, then $d$ forces $R_{e,i}$.)

6. For a condition $c = (k, \sigma_0, \ldots, \sigma_{k-1}, P)$, we say that *part $i$ of $c$ is acceptable* if there is an $X_0 \oplus \cdots \oplus X_{k-1} \in P$ such that $X_i \cap A$ and $X_i \cap \overline{A}$ are both infinite.

Definition A.3 below will give two ways in which we will extend conditions. We begin with an auxiliary definition.

**Definition A.2.** Let $P_0, \ldots, P_{m-1}$ be $k$-partition classes. Then $\bigotimes P_0, \ldots, P_{m-1}$ is the class of all sets of the form $\bigoplus_{i<k,\ a<b<m} X_i^a \cap X_i^b$ where $X_0^l \oplus \cdots \oplus X_{k-1}^l \in P_l$ for each $l < m$. We assume this join is arranged so that the sets of the form $X_0^a \cap X_0^b$ come first, followed by those of the form $X_1^a \cap X_1^b$, and so on.

Let $X_0^l \oplus \cdots \oplus X_{k-1}^l$ for $l < k + 1$ be $k$-partitions. We claim $\bigoplus_{i<k,\ a<b<k+1} X_i^a \cap X_i^b$ is a $k\binom{k+1}{2}$-partition. To see that this is the case, fix $n$, and for each $l < k+1$, let $i_l$ be such that $n \in X_{i_l}^l$. Since every $i_l$ is less than $k$, there must be $a < b < k+1$ such that $i_a = i_b$. Then $n \in X_{i_a}^a \cap X_{i_a}^b$. Thus, if $P_0, \ldots, P_k$ are $\Pi_1^{0,C}$ $k$-partition classes, then $\bigotimes P_0, \ldots, P_k$ is a $\Pi_1^{0,C}$ $k\binom{k+1}{2}$-partition class.

**Definition A.3.** Let $c = (k, \sigma_0, \ldots, \sigma_{k-1}, P)$ be a condition.

1. *Type 1 extensions.* Take an $X_0 \oplus \cdots \oplus X_{k-1} \in P$, a $Y$ satisfying the Mathias condition $(\sigma_i, X_i)$, and a $\tau$ such that $\sigma_i \preccurlyeq \tau \prec Y$. Let $Q$ be the class of all $Z_0 \oplus \cdots \oplus Z_{k-1} \in P$ such that $\tau$, thought of as a finite set, is a subset of $Z_i/\sigma_i$. Note that $Q$ is nonempty, as it contains $X_0 \oplus \cdots \oplus X_{k-1}$. Let $d = (k, \sigma_0, \ldots, \sigma_{i-1}, \tau, \sigma_{i+1}, \ldots, \sigma_{k-1}, Q)$. It is easy to check that $d$

is a condition, and that it extends $c$, with this extension being witnessed by the identity function.

A special case is when $Y = X_i/\sigma_i$. In this case, we could also take $Q$ to be the class of all $Z_0 \oplus \cdots \oplus Z_{k-1} \in P$ such that $\tau \prec Z_i/\sigma_i$.

2. *Type 2 extensions.* Let

$$\widehat{P} = \{Z_0 \oplus \cdots \oplus Z_{2k-1} :$$
$$(Z_0 \cup Z_1) \oplus (Z_2 \cup Z_3) \oplus \cdots \oplus (Z_{2k-2} \cup Z_{2k-1}) \in P\}.$$

That is, $\widehat{P}$ is the $\Pi_1^{0,C}$ class of all $2k$-partitions obtained from the $k$-partitions in $P$ by splitting each part into two (not necessarily disjoint) parts. Let $S_0, \ldots, S_{2k}$ be nonempty $\Pi_1^{0,C}$ subclasses of $\widehat{P}$ and let $Q = \bigotimes S_0, \ldots, S_{2k}$. As shown above, $Q$ is a $\Pi_1^{0,C}$ $2k\binom{2k+1}{2}$-partition class, so it is easy to check that

$$d = \left(2k\binom{2k+1}{2}, \sigma_0, \ldots, \sigma_0, \sigma_1, \ldots, \sigma_1, \ldots, \sigma_{k-1}, \ldots, \sigma_{k-1}, Q\right),$$

where each $\sigma_i$ appears $2\binom{2k+1}{2}$ many times, is a condition extending $c$.

**The general plan.** The proof will consist of establishing the following three lemmas; the proof of the third lemma is the core of the argument.

**Lemma A.4.** *Every condition has an acceptable part.*

**Lemma A.5.** *For every condition $c$ and every $m$, there is a condition $d$ extending $c$ such that $d$ forces $Q_m$ on each of its acceptable parts.*

**Lemma A.6.** *For every condition $c$ and every $e$ and $i$, there is a condition $d$ extending $c$ that forces $R_{e,i}$.*

Given these lemmas, it is easy to see that we can build a sequence of conditions $c_0, c_1, \ldots$ with the following properties.

1. Each $c_{s+1}$ extends $c_s$.
2. If $s = \langle e, i \rangle$ then $c_s$ forces $R_{e,i}$.
3. Each $c_s$ has an acceptable part.
4. If part $i$ of $c_s$ is acceptable, then $c_s$ forces $Q_s$ on part $i$.

Clearly, if part $j$ of $c_{s+1}$ refines part $i$ of $c_s$ and is acceptable, then part $i$ of $c_s$ is also acceptable. Thus we can think of the acceptable parts of our conditions as forming a tree under the refinement relation. This tree is finitely branching and infinite, so it has an infinite path. In other words, there are $i_0, i_1, \ldots$ such that for each $s$, part $i_{s+1}$ of $c_{s+1}$ refines part $i_s$

of $c_s$, and part $i_s$ of $c_s$ is acceptable, which implies that $c_s$ forces $Q_s$ on part $i_s$. Write $c_s = (k_s, \sigma_0^s, \ldots, \sigma_{k_s-1}^s, P_s)$. Let $G = \bigcup_s \sigma_{i_s}^s$. Let $U_s$ be the class of all $Y$ that satisfy $(\sigma_{i_s}^s, X_i)$ for some $X_0 \oplus \cdots \oplus X_{k_s-1} \in P_s$. Then $U_0 \supseteq U_1 \supseteq \cdots$, each $U_s$ contains an extension of $\sigma_{i_s}^s$, and the $U_s$ are all closed. Thus $G \in \bigcap_s U_s$. In other words, $G$ satisfies each $c_s$ on part $i_s$, and hence satisfies all of our requirements. Thus we are left with proving the lemmas.

*Proof of Lemma A.4.* It is here that we use the assumption that $A \not\leq_T C$. Let $c = (k, \sigma_0, \ldots, \sigma_{k-1}, P)$ be a condition. Since $P$ is a nonempty $\Pi_1^{0,C}$ class, the relativized form of the Cone Avoidance Basis Theorem 3.14 implies that there is an $X = X_0 \oplus \cdots \oplus X_{k-1} \in P$ such that $A \not\leq_T X$. If $c$ does not have an acceptable part then $A =^* \bigcup_{X_i \subseteq^* A} X_i$, so $A \leq_T X$, which is a contradiction. □

*Proof of Lemma A.5.* Fix $m$. It is enough to show that for a condition $c = (k, \sigma_0, \ldots, \sigma_{k-1}, P)$, if part $i$ of $c$ is acceptable, then there is a condition $d_0 = (k, \tau_0, \ldots, \tau_{k-1}, Q)$ extending $c$ such that $d_0$ forces $Q_m$ on part $i$, where the extension of $c$ by $d_0$ is witnessed by the identity map. (Note that if part $i$ of $d_0$ is acceptable, then so is part $i$ of $c$.) Then we can iterate this process, forcing $Q_m$ on each acceptable part in turn, to obtain the condition $d$ in the statement of the lemma.

So fix an acceptable part $i$ of $c$. Then there is a $\tau \succ \sigma_i$ for which there are $n_0, n_1 > m$ with $n_0 \in A$ and $n_1 \notin A$ such that $\tau(n_0) = \tau(n_1) = 1$, and there is an $X_0 \oplus \cdots \oplus X_{k-1} \in P$ with $\tau \prec X_i/\sigma_i$. Let $Q = \{X_0 \oplus \cdots \oplus X_{k-1} \in P : \tau \prec X_i/\sigma_i\}$. Let $d_0 = (k, \sigma_0, \ldots, \sigma_{i-1}, \tau, \sigma_{i+1}, \ldots, \sigma_{k-1}, Q)$. Then $d_0$ is a Type 1 extension of $c$, as in Definition A.3, and it clearly forces $Q_m$ on part $i$. □

*Proof of Lemma A.6.* Fix numbers $e$ and $i$, and a condition $c = (k, \sigma_0, \ldots, \sigma_{k-1}, P)$. For any condition $d$, let $U(d)$ be the set of all $j$ such that $d$ does not force $R_{e,i}$ on part $j$. If $U(c) = \emptyset$ then there is nothing to prove, so we assume $U(c) \neq \emptyset$. It is clearly enough to obtain a condition $d$ extending $c$ such that $|U(d)| < |U(c)|$. In fact, we will split the proof into two cases. In one, there will be a $j \in U(c)$ such that we can extend $\sigma_j$ to force $R_{e,i}$ via a Type 1 extension (as in Definition A.3). In the other, we will obtain $d$ via a Type 2 extension, and $d$ will in fact force $R_{e,i}$ (on all of its parts).

We will use the following notions. These may seem a bit mysterious at first, so before proceeding with the full proof, we will give an example illustrating the main idea of the proof. Here and below, we write $\sigma^A$ for the string of the same length as $\sigma$ defined by $\sigma^A(n) = 1$ iff $\sigma(n) = 1$ and $n \in A$, and similarly for $\sigma^{\overline{A}}$.

**Definition A.7.** 1. A *valuation* is a finite partial function $\mathbb{N} \to 2$.
2. A valuation $p$ is *correct* if $p(n) = \Phi_n(n)\!\downarrow$ for all $n \in \operatorname{dom} p$.
3. Valuations $p, q$ are *incompatible* if there is an $n$ such that $p(n) \neq q(n)$.
4. Let $c = (k, \sigma_0, \ldots, \sigma_{k-1}, P)$ be a condition and let $p$ be a valuation. We say that part $j$ of $c$ *disagrees* with $p$ if for every $X_0 \oplus \cdots \oplus X_{k-1} \in P$ and every $Z_0, Z_1$ with $X_j = Z_0 \cup Z_1$, there is a $Y$ such that

    (a) $Y$ satisfies the Mathias condition $(\sigma_j, X_j)$ and

    (b) there is an $n \in \operatorname{dom} p$ such that either $\Phi_e^{((Y \cap Z_0)/\sigma_j^A) \oplus C}(n)\!\downarrow\; \neq p(n)$
         or $\Phi_i^{((Y \cap Z_1)/\sigma_j^{\overline{A}}) \oplus C}(n)\!\downarrow\; \neq p(n)$.

To illustrate how we will use these notions, let us consider the simple case in which $c = (1, \lambda, P)$, where $\lambda$ is the empty string and $P$ is the unique 1-partition class, namely the class $\{\mathbb{N}\}$. Since $c$ has only one part, we will suppress references to parts of $c$ in our terminology. We will show in Lemma A.9 that either $c$ disagrees with some correct valuation, or there are three pairwise incompatible valuations $p_0, p_1, p_2$ such that $c$ does not disagree with any $p_j$ for $j < 2$.

In the former case, we argue as follows. By the definition of disagreeing with a correct valuation, applied to $Z_0 = A$ and $Z_1 = \overline{A}$, there are an $n$ and a $Y$ such that either $\Phi_e^{(Y \cap A) \oplus C}(n)\!\downarrow\; \neq \Phi_n(n)\!\downarrow$ or $\Phi_i^{(Y \cap \overline{A}) \oplus C}(n)\!\downarrow\; \neq \Phi_n(n)\!\downarrow$. If $\tau$ is a sufficiently long initial segment of $Y$, then for every $Z$ extending $\tau$, we have either $\Phi_e^{(Z \cap A) \oplus C}(n)\!\downarrow\; \neq \Phi_n(n)\!\downarrow$ or $\Phi_i^{(Z \cap \overline{A}) \oplus C}(n)\!\downarrow\; \neq \Phi_n(n)\!\downarrow$. Thus $(1, \tau, P)$ extends $c$ and forces $R_{e,i}$.

In the latter case, we argue as follows. For each $j < 3$, let $S_j$ be the class of all sets $Z_0 \oplus Z_1$ such that

1. $Z_0 \cup Z_1 = \mathbb{N}$.
2. if $Y_0 \subseteq Z_0$ then for every $n \in \operatorname{dom} p_j$, we have $\neg(\Phi_e^{Y_0 \oplus C}(n)\!\downarrow\; \neq p_j(n))$, and
3. if $Y_1 \subseteq Z_1$ then for every $n \in \operatorname{dom} p_j$, we have $\neg(\Phi_i^{Y_1 \oplus C}(n)\!\downarrow\; < p_j(n))$.

Since $c$ does not disagree with any of the $p_j$, the $S_j$ are nonempty. It is then easy to see that each $S_j$ is in fact a $\Pi_1^{0,C}$ 2-partition class.

Let $Q = \bigotimes S_0, S_1, S_2$ and let $d = (6, \lambda, \lambda, \lambda, \lambda, \lambda, \lambda, Q)$. Then $d$ is a Type 2 extension of $c$, as in Definition A.3.

Let $G$ satisfy $d$. Then there are some $a < b < 3$, some $Z_0 \oplus Z_1 \in S_a$, and some $W_0 \oplus W_1 \in S_b$ such that either $G \subseteq Z_0 \cap W_0$ or $G \subseteq Z_1 \cap W_1$. Let us suppose the former case holds, the latter case being similar. Then $G \cap A$ is a subset of both $Z_0$ and $W_0$. Let $n$ be such that $p_a(n) \neq p_b(n)$. By the definitions of $S_a$ and $S_b$, we have $\neg(\Phi_e^{(G \cap A) \oplus C}(n) \downarrow \neq p_a(n))$ and $\neg(\Phi_e^{(G \cap A) \oplus C}(n) \downarrow \neq p_b(n))$. Hence we must have $\Phi_e^{(G \cap A) \oplus C}(n) \uparrow$. Thus, $d$ forces $R_{e,i}$.

Let us now return to the general case. The following argument shows that we can give an alternate characterization of the notion of disagreement that highlights its effectiveness (relative to $C$).

**Lemma A.8.** *For each $j < k$, the set of all valuations $p$ such that part $j$ of $c$ disagrees with $p$ is $C$-c.e.*

*Proof.* For $\sigma \prec \tau$, say that $\nu$ *satisfies* $(\sigma, \tau)$ if $\sigma \preccurlyeq \nu$ and $|\nu| = |\tau|$, and $\nu(n) = 1 \Rightarrow \tau(n) = 1$ for all $n < |\nu|$. Write $\nu = \nu_0 \cup \nu_1$ to mean that all three strings have the same lengths, and $\nu(n) = 1 \Leftrightarrow \nu_0(n) = 1 \vee \nu_1(n) = 1$ for all $n < |\nu|$. Write $\nu = \nu_0 \cap \nu_1$ to mean that all three strings have the same length, and $\nu(n) = 1 \Leftrightarrow \nu_0(n) = 1 \wedge \nu_1(n) = 1$ for all $n < |\nu|$. If $\sigma$ is a string then $\Phi_e^{\sigma \oplus C}(n) \downarrow$ means that, thinking of $\sigma$ as a finite set, this computation converges with use at most $|\sigma|$.

Let $T$ be a $C$-computable tree such that $P$ is the set of infinite paths of $T$. For a valuation $p$, let $S_p$ be the subtree of $T$ defined as the closure under prefixes of the set of all $\tau_0 \oplus \cdots \oplus \tau_{k-1} \in T$ for which there exist $\rho_0, \rho_1$ with $\tau_j = \rho_0 \cup \rho_1$, such that for every $\nu$ satisfying $(\sigma_j, \tau_j)$ and every $n \in \operatorname{dom} p$, neither

$$\Phi_e^{((\nu \cap \rho_0)/\sigma_j^A) \oplus C}(n)[|\nu|] \downarrow \neq p(n) \qquad \text{nor} \qquad \Phi_i^{((\nu \cap \rho_1)/\sigma_j^{\overline{A}}) \oplus C}(n)[|\nu|] \downarrow \neq p(n).$$

We now show that part $j$ of $c$ disagrees with $p$ iff $S_p$ is finite.

First suppose $S_p$ is finite and let $X = X_0 \oplus \cdots \oplus X_{k-1} \in P$. Then there is a $\tau = \tau_0 \oplus \cdots \oplus \tau_{k-1}$ such that $\tau \prec X$ and $\tau \notin S_p$. Thus for all $\rho_0, \rho_1$ such that $\tau_j = \rho_0 \cup \rho_1$, there are a $\nu$ satisfying $(\sigma_j, \tau_j)$ and an $n \in \operatorname{dom} p$ such that either

$$\Phi_e^{((\nu \cap \rho_0)/\sigma_j^A) \oplus C}(n) \downarrow \neq p(n) \qquad \text{or} \qquad \Phi_i^{((\nu \cap \rho_1)/\sigma_j^{\overline{A}}) \oplus C}(n) \downarrow \neq p(n).$$

Let $Z_0$ and $Z_1$ be such that $X_j = Z_0 \cup Z_1$ and, letting $\rho_i = Z_i \upharpoonright |\tau_j|$, let $Y \subseteq X_j/\sigma_j$ extend a $\nu$ as above. Then $Y$ satisfies the Mathias condition $(\sigma_j, X_j)$, and there is an $n \in \operatorname{dom} p$ for which either

$$\Phi_e^{((Y \cap Z_0)/\sigma_j^A) \oplus C}(n) \downarrow \neq p(n) \qquad \text{or} \qquad \Phi_i^{((Y \cap Z_1)/\sigma_j^{\overline{A}}) \oplus C}(n) \downarrow \neq p(n).$$

Thus part $j$ of $c$ disagrees with $p$.

Now suppose $S_p$ is infinite and let $X_0 \oplus \cdots \oplus X_{k-1}$ be an infinite path of $S_p$. Let $Y$ be any set satisfying the Mathias condition $(\sigma_j, X_j)$. Let $U$ be the set of all strings $\rho_0 \oplus \rho_1$ such that there is an $m$ for which $\rho_0 \cup \rho_1 = X_j \restriction m$ and for each $n \in \operatorname{dom} p$, neither

$$\Phi_e^{((Y \restriction m \cap \rho_0)/\sigma_j^A) \oplus C}(n)[m]{\downarrow} \neq p(n) \quad \text{nor} \quad \Phi_i^{((Y \restriction m \cap \rho_1)/\sigma_j^{\overline{A}}) \oplus C}(n)[m]{\downarrow} \neq p(n).$$

Then the downward closure of $U$ (under string extension) is an infinite tree, and hence has an infinite path $Z_0 \oplus Z_1$. Then $X_j = Z_0 \cup Z_1$, and for each $n \in \operatorname{dom} p$, neither

$$\Phi_e^{((Y \cap Z_0)/\sigma_j^A) \oplus C}(n){\downarrow} \neq p(n) \quad \text{nor} \quad \Phi_i^{((Y \cap Z_1)/\sigma_j^{\overline{A}}) \oplus C}(n){\downarrow} \neq p(n).$$

Thus part $j$ of $c$ does not disagree with $p$.

So we see that part $j$ of $c$ disagrees with $p$ iff $S_p$ is finite. Since the $S_p$ are uniformly $C$-computable, the lemma follows. $\qquad\square$

The following lemma gives us two cases to consider. It is the one place in the construction where we use the fact that $C$ does not have PA degree. Recall that $U(c)$ is the set of all $j$ such that $c$ does not force $R_{e,i}$ on part $j$.

**Lemma A.9.** *One of the following must hold.*

1. *There is a correct valuation $p$ such that for some $j \in U(c)$, part $j$ of $c$ disagrees with $p$.*
2. *There are pairwise incompatible valuations $p_0, \ldots, p_{2k}$ such that for all $l < 2k + 1$ and all $j \in U(c)$, part $j$ of $c$ does not disagree with $p_l$.*

*Proof.* Assume that alternative 1 above does not hold. Since $C$ does not have PA degree, there is no $C$-computable function $h$ such that if $\Phi_n(n){\downarrow}$ then $h(n) = \Phi_n(n)$. Let $E$ be the set of all valuations $p$ such that for some $j \in U(c)$, part $j$ of $c$ disagrees with $p$. By Lemma A.8, $E$ is $C$-c.e.

Let $S$ be the collection of all finite sets $F$ such that for each $n \notin F$, either $\Phi_n(n){\downarrow}$ or there is a $p \in E$ such that $F \cup \{n\} \subseteq \operatorname{dom} p$ and for every $m \in \operatorname{dom} p \setminus F \cup \{n\}$, we have $p(m) = \Phi_m(m){\downarrow}$. If $F \notin S$, then there is at least one $n \notin F$ for which the above does not hold. We say that any such $n$ *witnesses* that $F \notin S$.

First suppose that $\emptyset \in S$. Then for each $n$, either $\Phi_n(n){\downarrow}$ or there is a $p \in E$ such that $n \in \operatorname{dom} p$ and for every $m \neq n$ in $\operatorname{dom} p$, we have $p(m) = \Phi_m(m){\downarrow}$. Then we can define $h \leqslant_{\mathrm{T}} C$ by waiting until either $\Phi_n(n){\downarrow}$, in which case we let $h(n) = \Phi_n(n)$, or a $p$ as above enters $E$, in which case we let $h(n) = 1 - p(n)$. Since no element of $E$ is correct, in the latter case, if $\Phi_n(n)\downarrow$ then $p(n) \neq \Phi_n(n)$, so $h(n) = \Phi_n(n)$. Since $C$ does not have PA degree, this case cannot occur.

Thus $\emptyset \notin S$. Let $n_0$ witness this fact. Given $n_0, \ldots, n_j$, if $\{n_0, \ldots, n_j\} \notin S$, then let $n_{j+1}$ witness this fact. Note that if $n_j$ is defined then $\Phi_{n_j}(n_j)\uparrow$.

Suppose that for some $j$, we have $\{n_0, \ldots, n_j\} \in S$. Then $\{n_0, \ldots, n_{j-1}\} \notin S$, as otherwise $n_j$ would not be defined. We define $h \leqslant_T C$ as follows. First, let $h(n_l) = 0$ for $l \leqslant j$. Given $n \notin \{n_0, \ldots, n_j\}$, we wait until either $\Phi_n(n)\downarrow$, in which case we let $h(n) = \Phi_n(n)$, or a $p$ enters $E$ such that $\{n_0, \ldots, n_j, n\} \subseteq \operatorname{dom} p$ and for every $m \in \operatorname{dom} p \setminus \{n_0, \ldots, n_j, n\}$, we have $p(m) = \Phi_m(m)\downarrow$. If $\Phi_n(n)\uparrow$ then the latter case must occur, since $\{n_0, \ldots, n_j\} \in S$. In this case, we cannot have $p(n) = \Phi_n(n)\downarrow$, as then $p$ would be a counterexample to the fact that $n_j$ witnesses that $\{n_0, \ldots, n_{j-1}\} \notin S$. Thus we can let $h(n) = 1 - p(n)$. Again, since $C$ does not have PA degree, this case cannot occur.

Thus $\{n_0, \ldots, n_j\} \notin S$ for all $j$. There are $2^{j+1}$ many valuations with domain $\{n_0, \ldots, n_j\}$, and they are all pairwise incompatible. None of these valuations can be in $E$, as that would contradict the fact that $n_j$ witnesses that $\{n_0, \ldots, n_{j-1}\} \notin S$. Taking $j$ large enough, we have $2k + 1$ many pairwise incompatible valuations, none of which are in $E$. $\qquad\square$

**Lemma A.10.** *If alternative 1 in Lemma A.9 holds, then $c$ has an extension $d$ such that $|U(d)| < |U(c)|$.*

*Proof.* Let $j \in U(c)$ be such that part $j$ of $c$ disagrees with some correct valuation. Let $X_0 \oplus \cdots \oplus X_{k-1} \in P$. Let $Z_0 = X_j \cap A$ and $Z_1 = X_j \cap \overline{A}$. By the definition of disagreeing with a correct valuation, there are an $n$ and a $Y$ satisfying the Mathias condition $(\sigma_j, X_j)$ such that either

$$\Phi_e^{((Y \cap Z_0)/\sigma_j^A)\oplus C}(n)\downarrow \neq \Phi_n(n)\downarrow \quad \text{or} \quad \Phi_i^{((Y \cap Z_1)/\sigma_j^{\overline{A}})\oplus C}(n)\downarrow \neq \Phi_n(n)\downarrow.$$

In other words, either

$$\Phi_e^{(Y \cap A)\oplus C}(n)\downarrow \neq \Phi_n(n)\downarrow \quad \text{or} \quad \Phi_i^{(Y \cap \overline{A})\oplus C}(n)\downarrow \neq \Phi_n(n)\downarrow.$$

If $\tau$ is a sufficiently long initial segment of $Y$, then for every $Z$ extending $\tau$, we have either

$$\Phi_e^{(Z \cap A)\oplus C}(n)\downarrow \neq \Phi_n(n)\downarrow \quad \text{or} \quad \Phi_i^{(Z \cap \overline{A})\oplus C}(n)\downarrow \neq \Phi_n(n)\downarrow.$$

We may assume that $\tau \succcurlyeq \sigma_j$. Let $Q$ be the class of all $W_0 \oplus \cdots \oplus W_{k-1} \in P$ such that $\tau$, thought of as a finite set, is a subset of $W_j/\sigma_j$ and let

$d = (k, \sigma_0, \ldots, \sigma_{j-1}, \tau, \sigma_{j+1}, \ldots, \sigma_{k-1}, Q)$. Then $d$ is a Type 1 extension of $c$, as in Definition A.3, and clearly $d$ forces $R_{e,i}$ on part $j$, so that $|U(d)| < |U(c)|$. $\qquad\square$

**Lemma A.11.** *If alternative 2 in Lemma A.9 holds, then $c$ has an extension $d$ that forces $R_{e,i}$.*

*Proof.* Let $p_0, \ldots, p_{2k}$ be pairwise incompatible valuations such that for all $l < 2k + 1$ and all $j \in U(c)$, part $j$ of $c$ does not disagree with $p_l$. For each $l < 2k + 1$, let $S_l$ be the class of all sets of the form $Z_0 \oplus \cdots \oplus Z_{2k-1}$ such that

1. $(Z_0 \cup Z_1) \oplus (Z_2 \cup Z_3) \oplus \cdots \oplus (Z_{2k-2} \cup Z_{2k-1}) \in P$,
2. if $Y$ satisfies the Mathias condition $(\sigma_j, Z_{2j})$ and $j \in U(c)$ then for every $n \in \operatorname{dom} p_l$, we have $\neg(\Phi_e^{Y/\sigma_j^A \oplus C}(n){\downarrow} \neq p_l(n))$, and
3. if $Y$ satisfies the Mathias condition $(\sigma_j, Z_{2j+1})$ and $j \in U(c)$ then for every $n \in \operatorname{dom} p_l$, we have $\neg(\Phi_i^{Y/\sigma_j^{\overline{A}} \oplus C}(n){\downarrow} \neq p_l(n))$.

Since for every $j \in U(c)$, part $j$ of $c$ does not disagree with any of the $p_l$, the $S_l$ are nonempty. It is then easy to see that each $S_l$ is in fact a $\Pi_1^{0,C}$ $2k$-partition class.

Let $Q = \bigotimes S_0, \ldots, S_{2k}$ and let

$$d = \left(2k\binom{2k+1}{2}, \sigma_0, \ldots, \sigma_0, \sigma_1, \ldots, \sigma_1, \ldots, \sigma_{k-1}, \ldots, \sigma_{k-1}, Q\right),$$

where each $\sigma_i$ appears $2\binom{2k+1}{2}$ many times. Then $d$ is a Type 2 extension of $c$, as in Definition A.3.

Let $G$ satisfy $d$. Then there are some $j < k$, some $a < b < 2k + 1$, some $Z_0 \oplus \cdots \oplus Z_{2k-1} \in S_a$, and some $W_0 \oplus \cdots \oplus W_{2k-1} \in S_b$ such that $G$ satisfies one of the Mathias conditions $(\sigma_j, Z_{2j} \cap W_{2j})$ or $(\sigma_j, Z_{2j+1} \cap W_{2j+1})$. Then $G$ satisfies $c$ on part $j$, so if $j \notin U(c)$, then $G$ satisfies $R_{e,i}$. So assume $j \in U(c)$.

Let us suppose $G$ satisfies $(\sigma_j, Z_{2j} \cap W_{2j})$, the other case being similar. Then $(G \cap A)/\sigma_j$ satisfies both of the Mathias conditions $(\sigma_j, Z_{2j})$ and $(\sigma_j, W_{2j})$. Let $n$ be such that $p_a(n) \neq p_b(n)$. By the definitions of $S_a$ and $S_b$, we have $\neg(\Phi_e^{(G \cap A) \oplus C}(n){\downarrow} \neq p_a(n))$ and $\neg(\Phi_e^{(G \cap A) \oplus C}(n){\downarrow} \neq p_b(n))$. Hence we must have $\Phi_e^{(G \cap A) \oplus C}(n){\uparrow}$. Thus $d$ forces $R_{e,i}$. $\qquad\square$

The previous three lemmas together complete the proof. $\qquad\square$

# Bibliography

[1] Computability, reverse mathematics and combinatorics: open problems, `http://www.birs.ca/workshops/2008/08w5019/files/problems.pdf`.

[2] K. Ambos-Spies, B. Kjos-Hanssen, S. Lempp, and T. A. Slaman, Comparing DNR and WWKL, J. Symbolic Logic 69 (2004) 1089–1104.

[3] C. J. Ash and J. F. Knight, Computable Structures and the Hyperarithmetical Hierarchy, Stud. Logic Found. Math. 144, North-Holland, Amsterdam, 2000.

[4] A. Blass, Some questions arising from Hindman's Theorem, Sci. Math. Jpn. 62 (2005) 331–334.

[5] A. R. Blass, J. L. Hirst, and S. G. Simpson, Logical analysis of some theorems of combinatorics and topological dynamics, in S. G. Simpson, ed., Logic and Combinatorics, Contemp. Math. 65, Amer. Math. Soc., Providence, RI, 1987, 125–156.

[6] R. Bonnet, Stratifications et extension des genres de chaînes dénombrables, C. R. Acad. Sci Paris Sér. A-B 269 (1969) A880–A882 (in French).

[7] R. Bonnet and M. Pouzet, Linear extensions of ordered sets, in I. Rival, ed., Ordered Sets, D. Reidel Publishing Co., Dordrecht-Boston, 1982, 125–170.

[8] W. W. Boone, The word problem, Proc. Nat. Acad. Sci. U.S.A. 44 (1958) 1061–1065.

[9] A. Bovykin and A. Weiermann, The strength of infinitary Ramseyan principles can be accessed by their densities, to appear in Ann. Pure Appl. Logic.

[10] V. Brattka and G. Gherardi, Effective choice and boundedness principles in computable analysis, Bull. Symbolic Logic 17 (2011) 73–117.

[11] V. Brattka and G. Gherardi, Weihrauch degrees, omniscience principles and weak computability, J. Symbolic Logic 76 (2011) 143–176.

[12] D. Bridges and E. Palmgren, Constructive mathematics, in E. N. Zalta, ed., The Stanford Encyclopedia of Philosophy, Winter 2013 edition, `http://plato.stanford.edu/archives/win2013/entries/mathematics-constructive/`.

[13] T. J. Carlson, Some unifying principles in Ramsey theory, Discrete Math. 68 (1988) 117–169.

[14] D. Cenzer and C. G. Jockusch, Jr., $\Pi_1^0$ classes: structure and applications, in P. A. Cholak, S. Lempp, M. Lerman, and R. A. Shore, eds., Computability Theory and its Applications, Contemp. Math. 257, Amer. Math. Soc., Providence, RI, 2000, 39–59.

[15] D. Cenzer and J. B. Remmel, $\Pi_1^0$ classes in mathematics, in [55], vol. 2, 623–822.

[16] D. Cenzer and J. B. Remmel, Effectively Closed Sets, to be published in ASL Lecture Notes in Logic, draft version available at time of writing at http://people.clas.ufl.edu/cenzer/files/book4.pdf.

[17] C. C. Chang and H. J. Keisler, Model Theory, Third edition, Stud. Logic Found. Math. 73 North-Holland, Amsterdam, 1990.

[18] P. A. Cholak, D. D. Dzhafarov, and J. L. Hirst, On Mathias generic sets, in S. B. Cooper, A. Dawar, and B. Löwe, eds., How the World Computes, Lecture Notes in Computer Science 7318, Springer-Verlag, Berlin, 2012, 129–138.

[19] P. A. Cholak, M. Giusto, J. L. Hirst, and C. G. Jockusch, Jr., Free sets and reverse mathematics, in [189] 104–119.

[20] P. A. Cholak, C. G. Jockusch, Jr., and T. A. Slaman, On the strength of Ramsey's Theorem for pairs, J. Symbolic Logic 66 (2001) 1–55 (corrigendum in J. Symbolic Logic 74 (2009) 1438–1439).

[21] P. Cholak, A. Marcone, and R. Solomon, Reverse mathematics and the equivalence of definitions for well and better quasi-orders. J. Symbolic Logic 69 (2004) 683–712.

[22] C. T. Chong, S. Lempp, and Y. Yang, On the role of the collection principle for $\Sigma_2^0$-formulas in second-order reverse mathematics, Proc. Amer. Math. Soc. 138 (2010) 1093-1100

[23] C. T. Chong, T. A. Slaman, and Y. Yang, $\Pi_1^1$-conservation of combinatorial principles weaker than Ramsey's Theorem for pairs, Adv. Math. 230 (2012) 1060–1077.

[24] C. T. Chong, T. A. Slaman, and Y. Yang, The metamathematics of stable Ramsey's Theorem for pairs, J. Amer. Math. Soc. 27 (2014) 863–892.

[25] C. T. Chong and Y. Yang, Computability theory in arithmetic: provability, structure, and techniques, in P. A. Cholak, S. Lempp, M. Lerman, and R. A. Shore, eds., Computability Theory and its Applications, Boulder, CO, 1999, Contemp. Math. 257, Amer. Math. Soc., Providence, RI, 2000, 73–81.

[26] C. T. Chong and Y. Yang, Recursion theory on weak fragments of Peano arithmetic: a study of definable cuts, in C. T. Chong, Q. Feng, D. Ding, Q. Huang, and M. Yasugi, eds., Proceedings of the Sixth Asian Logic Conference, Beijing, 1996, World Sci. Publ., River Edge, NJ, 1998, 47–65.

[27] J. Chubb, J. L. Hirst, and T. H. McNicholl, Reverse mathematics, computability, and partitions of trees, J. Symbolic Logic 74 (2009) 201–215.

[28] C. J. Conidis, Classifying model-theoretic properties, J. Symbolic Logic 73 (2008) 885–905.

[29] C. J. Conidis and T. A. Slaman, Random reals, the rainbow Ramsey the-

orem, and arithmetic conservation, J. Symbolic Logic 78 (2013) 195–206.

[30] J. Corduan, M. J. Groszek, and J. R. Mileti, Reverse mathematics and Ramsey's property for trees, J. Symbolic Logic 75 (2010) 945–954.

[31] B. F. Csima, Degree spectra of prime models, J. Symbolic Logic 69 (2004) 430–442.

[32] B. F. Csima, D. R. Hirschfeldt, V. S. Harizanov, and R. I. Soare, Bounding homogeneous models, J. Symbolic Logic 72 (2007) 305–323.

[33] B. F. Csima, D. R. Hirschfeldt, J. F. Knight, and R. I. Soare, Bounding prime models, J. Symbolic Logic 69 (2004) 1117–1142.

[34] B. F. Csima and J. R. Mileti, The strength of the rainbow Ramsey theorem, J. Symbolic Logic 74 (2009) 1310–1324.

[35] D. H. J. de Jongh and R. Parikh, Well-partial orderings and hierarchies, Nederl. Akad. Wetensch. Proc. Ser. A 80 = Indag. Math. 39 (1977) 195–207.

[36] J. C. E. Dekker and J. Myhill, Recursive equivalence types, Univ. California Publ. Math. 3 (1960) 67–213.

[37] D. Diamondstone, R. Downey, N. Greenberg, and D. Turetsky, The finite intersection principle and genericity, to appear, available at time of writing at http://homepages.ecs.vuw.ac.nz/~downey/publications/FIP_paper.pdf.

[38] F. G. Dorais, D. D. Dzhafarov, J. L. Hirst, J. R. Mileti, and P. Shafer, On uniform relationships between combinatorial problems, to appear in Trans. Amer. Math. Soc., available at http://arxiv.org/abs/1212.0157.

[39] R. G. Downey, Computability theory and linear orderings, in [55], vol. 2, 823–976.

[40] R. G. Downey and D. R. Hirschfeldt, Algorithmic Randomness and Complexity, Theory and Applications of Computability, Springer, New York, 2010.

[41] R. G. Downey, D. R. Hirschfeldt, S. Lempp, and R. Solomon, A $\Delta_2^0$ set with no infinite low subset in either it or its complement, J. Symbolic Logic 66 (2001) 1371–1381.

[42] R. G. Downey, D. R. Hirschfeldt, S. Lempp, and R. Solomon, Computability-theoretic and proof-theoretic aspects of partial and linear orderings, Israel J. Math. 138 (2003) 271–290.

[43] R. G. Downey, C. G. Jockusch, Jr., and J. S. Miller, On self-embeddings of computable linear orderings, Ann. Pure Appl. Logic 138 (2006) 52–76.

[44] R. G. Downey and S. Lempp, The proof-theoretic strength of the Dushnik-Miller theorem for countable linear orders, in M. M. Arslanov and S. Lempp, eds., Recursion Theory and Complexity, de Gruyter Ser. Log. Appl. 2, de Gruyter, Berlin, 1999, 55–57.

[45] D. D. Dzhafarov, Stable Ramsey's theorem and measure, Notre Dame J. Formal Logic 52 (2011) 95–112.

[46] D. D. Dzhafarov, Infinite saturated orders, Order 28 (2011), 163–172.

[47] D. D. Dzhafarov, Cohesive avoidance and strong reductions, to appear in Proc. Amer. Math. Soc., available at http://arxiv.org/abs/1212.0828.

[48] D. D. Dzhafarov and J. L. Hirst, The polarized Ramsey's theorem, Arch. Math. Logic 48 (2009) 141–157.

[49] D. D. Dzhafarov, J. L. Hirst, and T. J. Lakins, Ramsey's theorem for trees: the polarized tree theorem and notions of stability, Arch. Math. Logic 49 (2010) 399–415.

[50] D. D. Dzhafarov and C. G. Jockusch, Jr., Ramsey's Theorem and cone avoidance, J. Symbolic Logic 74 (2009) 557–578.

[51] D. D. Dzhafarov and C. Mummert, Reverse mathematics and properties of finite character, Ann. Pure Appl. Logic 163 (2012) 1243–1251.

[52] D. D. Dzhafarov and C. Mummert, On the strength of the finite intersection principle, Israel J. Math. 196 (2013) 345–361.

[53] H. B. Enderton, A Mathematical Introduction to Logic, Second edition, Harcourt/Academic Press, Burlington, MA, 2001.

[54] P. Erdős, A. Hajnal, and R. Rado. Partition relations for cardinal numbers, Acta Math. Acad. Sci. Hungar. 16 (1965) 93–196.

[55] Yu. L. Ershov, S. S. Goncharov, A. Nerode, J. B. Remmel, and V. W. Marek, eds., Handbook of Recursive Mathematics Vols. I and II, Studies in Logic and the Foundations of Mathematics 138–139, North-Holland, Amsterdam, 1998.

[56] S. Flood, Reverse mathematics and a Ramsey-type König's Lemma, J. Symbolic Logic 77 (2012) 1272–1280.

[57] R. M. Friedberg, Two recursively enumerable sets of incomparable degrees of unsolvability, Proc. Nat. Acad. Sci. U.S.A. 43 (1957) 236–238.

[58] R. M. Friedberg, A criterion for completeness of degrees of unsolvability, J. Symbolic Logic 22 (1957) 159–160.

[59] H. Friedman, Some systems of second order arithmetic and their use, in R. D. James, ed., Proceedings of the International Congress of Mathematicians, Vancouver 1974, vol. 1, Canadian Mathematical Congress, 1975, 235–242.

[60] H. Friedman, Systems of second order arithmetic with restricted induction I, II (abstracts), J. Symbolic Logic 41 (1976) 557–559.

[61] H. M. Friedman, postings to the Foundations of Mathematics mailing list, http://www.cs.nyu.edu/pipermail/fom/1999-July/003257.html and http://www.cs.nyu.edu/pipermail/fom/1999-July/003263.html.

[62] H. M. Friedman. The inevitability of logical strength: strict reverse mathematics, in S. B. Cooper, H. Geuvers, A. Pillay, and J. Väänänen. Logic Colloquium 2006, Lect. Notes Log. 32, Assoc. Symbol. Logic, Chicago, IL, 2009, 135–183.

[63] H. M. Friedman and J. L. Hirst, Weak comparability of well orderings and reverse mathematics, Ann. Pure Appl. Logic 47 (1990) 11–29.

[64] H. M. Friedman, K. McAloon, and S. G. Simpson, A finite combinatorial principle which is equivalent to the 1-consistency of predicative analysis, Patras Logic Symposion, Stud. Logic Foundations Math. 109, North-Holland, Amsterdam-New York, 1982, 197–230.

[65] H. Friedman, N. Robertson, and P. Seymour, The metamathematics of the graph minor theorem, in S. G. Simpson, ed., Logic and Combinatorics, Contemp. Math. 65, Amer. Math. Soc., Providence, RI, 1987, 229–261.

[66] H. M. Friedman and S. G. Simpson, Issues and problems in reverse mathematics, in P. A. Cholak, S. Lempp, M. Lerman, and R. A. Shore, eds., Computability Theory and its Applications, Contemp. Math. 257, Amer. Math. Soc., Providence, RI, 2000, 127–144.

[67] H. M. Friedman, S. G. Simpson, and R. L. Smith, Countable algebra and set existence axioms, Ann. Pure Appl. Logic 25 (1983) 141–181.

[68] J. H. Gallier, What's so special about Kruskal's theorem and the ordinal $\Gamma_0$? A survey of some results in proof theory, Ann. Pure Appl. Logic 53 (1991) 199–260.

[69] R. O. Gandy, G. Kreisel, and W. W. Tait, Set existence, Bull. Acad. Polon. Sci. Sér. Sci. Math. Astronom. Phys. 8 (1960) 577–582.

[70] W. Gasarch, A survey of recursive combinatorics, in [55], vol. 2, 1041–1176.

[71] M. Giusto and S. G. Simpson, Located sets and reverse mathematics, J. Symbolic Logic 65 (2000) 1451–1480.

[72] S. S. Goncharov and A. T. Nurtazin, Constructive models of complete decidable theories, Algebra Logic 12 (1973) 67–77.

[73] N. Greenberg, J. D. Hamkins, D. R. Hirschfeldt, and R. Miller, eds., Effective Mathematics of the Uncountable, Lecture Notes in Logic 41, Association for Symbolic Logic, La Jolla, CA and Cambridge University Press, Cambridge, 2013.

[74] P. Hájek, Interpretability and fragments of arithmetic, in P. Clote and K. Krajíček, eds., Arithmetic, Proof Theory, and Computational Complexity (Prague, 1991), Oxford Logic Guides 23, Oxford Univ. Press, New York, 1993, 185–196.

[75] P. Hájek and P. Pudlák, Metamathematics of First-order Arithmetic, second printing, Perspect. Math. Logic, Springer-Verlag, Berlin, 1998.

[76] W. P. Hanf, The Boolean algebra of logic, Bull. Amer. Math. Soc. 81 (1975) 587–589.

[77] V. S. Harizanov, Pure computable model theory, in [55], vol. 1, 3–114.

[78] V. S. Harizanov, Turing degrees of certain isomorphic images of computable relations, Ann. Pure Appl. Logic 93 (1998) 103–113.

[79] L. Harrington, Recursively presentable prime models, J. Symbolic Logic 39 (1974) 305–309.

[80] K. Harris, The Complexity of Classical Theorems on Saturated Models, PhD Dissertation, The University of Chicago, 2007.

[81] K. Harris, On bounding saturated models, preprint, available at time of writing at http://people.cs.uchicago.edu/~kaharris/papers/computability/sat.pdf.

[82] E. Herrmann, Infinite chains and antichains in computable partial orderings, J. Symbolic Logic 66 (2001) 923–934.

[83] N. Hindman. Finite sums from sequences within cells of a partition of $N$. J. Combinatorial Theory Ser. A 17 (1974) 1–11.

[84] D. R. Hirschfeldt, Computable trees, prime models, and relative decidability, Proc. Amer. Math. Soc. 134 (2006) 1495–1498.

[85] D. R. Hirschfeldt and C. G. Jockusch, Jr., On notions of computability theoretic reduction between $\Pi^1_2$ principles, to appear, when available will

be found at `http://math.uchicago.edu/~drh/papers.html`.

[86]  D. R. Hirschfeldt, C. G. Jockusch, Jr., B. Kjos-Hanssen, S. Lempp, and T. A. Slaman, The strength of some combinatorial principles related to Ramsey's Theorem for pairs, in C. Chong, Q. Feng, T. A. Slaman, W. H. Woodin, and Y. Yang, eds., Computational Prospects of Infinity, Part II: Presented Talks, Lecture Notes Series, Institute for Mathematical Sciences, National University of Singapore, vol. 15, World Scientific, Singapore, 2008, 143–146.

[87]  D. R. Hirschfeldt, K. Lange, and R. A. Shore, Induction, bounding, weak combinatorial principles, and the Homogeneous Model Theorem, to appear, available at `http://math.uchicago.edu/~drh/Papers/hmt.html`.

[88]  D. R. Hirschfeldt and R. A. Shore, Combinatorial principles weaker than Ramsey's Theorem for pairs, J. Symbolic Logic 72 (2007) 171–206.

[89]  D. R. Hirschfeldt, R. A. Shore, and T. A. Slaman, The Atomic Model Theorem and type omitting, Trans. Amer. Math. Soc. 361 (2009) 5805–5837.

[90]  J. L. Hirst, Combinatorics in Subsystems of Second Order Arithmetic, PhD Dissertation, The Pennsylvania State University, 1987.

[91]  J. L. Hirst, Hindman's Theorem, ultrafilters, and reverse mathematics, J. Symbolic Logic 69 (2004) 65–72.

[92]  J. L. Hirst, A survey of the reverse mathematics of ordinal arithmetic, in [189] 222–234.

[93]  J. L. Hirst, Hilbert versus Hindman, Arch. Math. Logic 51 (2012) 123–125.

[94]  P. Howard and J. E. Rubin, Consequences of the Axiom of Choice, Mathematical Surveys and Monographs 59, American Mathematical Society, Providence, RI, 1998.

[95]  T. Hummel and C. G. Jockusch, Jr., Generalized cohesiveness, J. Symbolic Logic 64 (1999) 489–516.

[96]  C. G. Jockusch, Jr., The degrees of hyperhyperimmune sets, J. Symbolic Logic 34 (1969) 489–493.

[97]  C. G. Jockusch, Jr., Ramsey's Theorem and recursion theory, J. Symbolic Logic 37 (1972) 268–280.

[98]  C. G. Jockusch, Jr., Degrees in which the recursive sets are uniformly recursive, Canad. J. Math. 24 (1972) 1092–1099.

[99]  C. G. Jockusch, Jr., Upward closure and cohesive degrees, Israel J. Math. 15 (1973) 332–335.

[100]  C. G. Jockusch, Jr., Degrees of generic sets, in F. R. Drake and S. S. Wainer, eds., Recursion Theory: its Generalisations and Applications, London Math. Soc. Lecture Note Ser. 45, Cambridge Univ. Press, Cambridge, 1980, 110–139.

[101]  C. G. Jockusch, Jr., B. Kastermans, S. Lempp, M. Lerman, and R. Solomon, Stability and posets, J. Symbolic Logic 74 (2009) 693–711.

[102]  C. G. Jockusch, Jr., M. Lerman, R. I. Soare, and R. M. Solovay, Recursively enumerable sets modulo iterated jumps and extensions of Arslanov's completeness criterion, J. Symbolic Logic 54 (1989) 1288–1323.

[103]  C. G. Jockusch, Jr., A. A. Lewis, and J. B. Remmel, $\Pi_1^0$-classes and Rado's

selection principle, J. Symbolic Logic 56 (1991) 684–693.

[104] C. G. Jockusch, Jr. and T. G. McLaughlin, Countable retracing functions and $\Pi_2^0$ predicates, Pacific J. Math. 30 (1969) 67–93.

[105] C. G. Jockusch, Jr. and R. I. Soare, $\Pi_1^0$ classes and degrees of theories, Trans. Amer. Math. Soc. 173 (1972) 33–56.

[106] C. G. Jockusch, Jr. and R. I. Soare, Degrees of members of $\Pi_1^0$ classes, Pacific J. Math. 40 (1972) 605–616.

[107] C. G. Jockusch, Jr. and F. Stephan, A cohesive set which is not high, Math. Log. Quart. 39 (1993) 515–530 (correction in Math. Log. Quart. 43 (1997) 569).

[108] P. Jullien, Contribution á l'Étude des Types d'Ordre Dispersés, PhD Dissertation, Marseille, 1969 (in French).

[109] R. Kaye, Models of Peano Arithmetic, Oxford Logic Guides 15, Oxford University Press, New York, 1991.

[110] S. C. Kleene, On notation for ordinal numbers, J. Symbolic Logic 3 (1938) 150–155.

[111] S. C. Kleene and E. L. Post, The uppersemilattice of degrees of recursive unsolvability, Ann. Math. 59 (1954) 379–407.

[112] U. Kohlenbach, Higher order reverse mathematics, in [189] 281–295.

[113] D. König, Über eine Schlussweise aus dem Endlichen ins Undendliche, Acta Litt. Acad. Sci. Hung. (Szeged) 3 (1927) 121–130 (in German).

[114] G. Kreisel, A variant to Hilbert's theory of the foundations of arithmetic, British J. Philos. Sci. 4 (1953) 107–129.

[115] G. Kreisel, Analysis of the Cantor-Bendixson theorem by means of the analytic hierarchy, Bull. Acad. Polon. Sci. Sér. Sci. Math. Astronom. Phys. 7 (1959) 621–626.

[116] S. Kritchman and R. Raz, The surprise examination paradox and the second incompleteness theorem, Notices Amer. Math. Soc. 57 (Dec. 2010) 1454–1458.

[117] J. B. Kruskal, Well-quasi-ordering, the Tree Theorem, and Vazsonyi's conjecture, Trans. Amer. Math. Soc. 95 (1960) 210–225.

[118] K. Kunen, Set Theory. An Introduction to Independence Proofs, Stud. Logic Foundations Math. 102, North-Holland, Amsterdam-New York, 1980.

[119] D. Lacombe, Sur le semi-réseau constitué par lies degrés d'indecidabilité récursive, Compt. Rend. Ac. Sci. 239 (1954) 1108–1109 (in French).

[120] K. Lange, The degree spectra of homogeneous models, J. Symbolic Logic 73 (2008) 1009–1028.

[121] K. Lange, A characterization of the 0-basis homogeneous bounding degrees, J. Symbolic Logic 75 (2010) 971–995.

[122] K. Lange and R. Soare, Computability of homogeneous models, Notre Dame J. Formal Logic 48 (2007) 143–170.

[123] R. Laver, On Fraïssé's order type conjecture, Ann. Math. (2) 93 (1971) 89–111.

[124] M. Lerman, On recursive linear orderings, in M. Lerman, J. H. Schmerl, and R. I. Soare, eds., Logic Year 1979–1980, Lecture Notes in Math. 859, Springer-Verlag, Berlin, 1981, 132–142.

[125]  M. Lerman, R. Solomon, and H. Towsner, Separating principles below Ram-
       sey's Theorem for Pairs, J. Math. Log. 13 (2013) 1350007, 44 pp.
[126]  J. Liu, $RT_2^2$ does not imply $WKL_0$, J. Symbolic Logic 77 (2012) 609–620.
[127]  L. Liu, Cone avoiding closed sets, to appear in Trans. Amer. Math. Soc.
[128]  D. A. Martin, Classes of recursively enumerable sets and degrees of unsolv-
       ability. Z. Math. Logik Grundlag. Math. 12 (1966) 295–310.
[129]  A. Marcone, Wqo and bqo theory in subsystems of second order arithmetic,
       in [189] 303–330.
[130]  A. Marcone and A. Montalbán, On Fraïssé's conjecture for linear orders of
       finite Hausdorff rank, Ann. Pure Appl. Logic 160 (2009) 355–367.
[131]  A. Marcone and R. A. Shore, The maximal linear extension theorem in
       second order arithmetic, Arch. Math. Logic 50 (2011) 543–564.
[132]  D. Marker, Model Theory: An Introduction, Grad. Texts in Math. 277,
       Springer, New York, 2002.
[133]  K. McAloon, Paris-Harrington incompleteness and progressions of theories,
       Recursion Theory (Ithaca, N.Y., 1982), Proc. Sympos. Pure Math. 42,
       Amer. Math. Soc., Providence, RI, 1985, 447–460.
[134]  T. H. McNicholl, The Inclusion Problem for Generalized Frequency Classes,
       PhD Dissertation, The George Washington University, 1995.
[135]  Y. T. Medvedev, Degrees of difficulty of the mass problem, Dokl. Akad.
       Nauk SSSR (N.S.) 104 (1955) 501–504 (in Russian).
[136]  G. Metakides and A. Nerode, Effective content of field theory, Ann. Math.
       Logic 17 (1979) 289–320.
[137]  J. R. Mileti, Partition Theorems and Computability Theory, PhD Disser-
       tation, University of Illinois at Urbana-Champaign, 2004.
[138]  J. R. Mileti, The canonical Ramsey theorem and computability theory,
       Trans. Amer. Math. Soc. 360 (2008) 1309–1340.
[139]  T. S. Millar, The Theory of Recursively Presented Models, PhD Disserta-
       tion, Cornell University, 1976.
[140]  T. S. Millar, Foundations of recursive model theory, Ann. Math. Logic 13
       (1978) 45–72.
[141]  T. S. Millar, Omitting types, type spectrums, and decidability, J. Symbolic
       Logic 48 (1983) 171–181.
[142]  J. S. Miller, Assorted results in and about effective randomness, in prepa-
       ration.
[143]  J. S. Miller and R. Solomon, Effectiveness for infinite variable words and
       the Dual Ramsey Theorem, Arch. Math. Logic, 43 (2004) 543–555.
[144]  A. Montalbán, Equivalence between Fraïssé's Conjecture and Jullien's The-
       orem. Ann. Pure Appl. Logic 139 (2006) 1–42.
[145]  A. Montalbán, Computable linearizations of well-partial-orderings, Order
       24 (2007) 39–48.
[146]  A. Montalbán, Open questions in reverse mathematics, Bull. Symbolic
       Logic 17 (2011) 431–454.
[147]  A. Montalbán, Research statement, Oct. 2, 2011, available at time of writ-
       ing at http://math.berkeley.edu/~antonio/research4Log10.pdf.
[148]  A. Montalbán and R. A. Shore, The limits of determinacy in second order

arithmetic, Proc. London Math. Soc. 104 (2012) 223–252.

[149] M. Morley, Decidable models, Israel J. Math 25 (1976) 233–240.

[150] A. A. Muchnik, On the unsolvability of the problem of reducibility in the theory of algorithms, Dokl. Akad. Nauk SSSR (N.S.) 108 (1956) 194–197 (in Russian).

[151] A. A. Muchnik, On strong and weak reducibilities of algorithmic problems, Sib. Math. Zh. 4 (1963) 1328–1341 (in Russian).

[152] C. Mummert and S. G. Simpson, Reverse mathematics and $\Pi_2^1$ comprehension, Bull. Symbolic Logic 11 (2005) 526–533.

[153] M. E. Mytilinaios, Finite injury and $\Sigma_1$-induction, J. Symbolic Logic 54 (1989) 212–221.

[154] I. Neeman, The strength of Jullien's indecomposability theorem, J. Math. Log. 8 (2008) 93–119.

[155] I. Neeman, Necessary use of $\Sigma_1^1$ induction in a reversal, J. Symbolic Logic 76 (2011) 561–574.

[156] P. S. Novikov, On the algorithmic unsolvability of the word problem in group theory, Trudy Mat. Inst. im. Steklov. 44 (1955) 1–143 (in Russian).

[157] A. T. Nurtazin, Strong and weak constructivizations and enumerable families, Algebra and Logic 13 (1974) 177–184.

[158] P. Odifreddi, Classical Recursion Theory, Studies in Logic and the Foundations of Mathematics 125, North-Holland, Amsterdam, 1990.

[159] P. Odifreddi, Classical Recursion Theory Vol. II, Studies in Logic and the Foundations of Mathematics 143, North-Holland, Amsterdam, 1999.

[160] J. Paris and L. Harrington, A mathematical incompleteness in Peano Arithmetic, in J. Barwise, ed., Handbook of Mathematical Logic, North-Holland, Amsterdam, 1977, 1133–1142.

[161] J. B. Paris and L. A. S. Kirby, $\Sigma_n$-collection schemas in arithmetic, in Logic Colloquium '77, Stud. Logic Foundations Math. 96, North-Holland, Amsterdam-New York, 1978, 199–209.

[162] B. Poonen, Undecidability in number theory, Notices Amer. Math. Soc. 55 (Mar. 2008) 344–350.

[163] D. B. Posner and R. W. Robinson, Degrees joining to $\mathbf{0}'$, J. Symbolic Logic 46 (1981) 714–722.

[164] E. L. Post, Recursively enumerable sets of positive integers and their decision problems, Bull. Amer. Math. Soc. 50 (1944) 284–316.

[165] E. L. Post, Degrees of recursive unsolvability, Bull. Amer. Math. Soc. 54 (1948) 641–642.

[166] F. P. Ramsey, On a problem in formal logic, Proc. London Math. Soc. (3) 30 (1930) 264–286.

[167] M. Rathjen, The art of ordinal analysis, in M. Sanz-Solé, J. Soria, J. L. Varona, and J. Verdera, eds., International Congress of Mathematicians, vol. II, Eur. Math. Soc, Zürich, 2006, 45–69.

[168] M. Rathjen and A. Weiermann, Proof-theoretic investigations on Kruskal's theorem, Ann. Pure Appl. Logic 60 (1993) 49–88.

[169] B. Rice, The Thin Set Theorem for pairs implies DNR, to appear in Notre Dame J. Formal Logic, available at time of writing at http://www.math.

`wisc.edu/~brice/research.html`.

[170]  H. Rogers, Jr., Theory of Recursive Functions and Effective Computability, Second edition, MIT Press, Cambridge, MA, 1987.

[171]  G. F. Rose and J. S. Ullian, Approximations of functions on the integers, Pacific J. Math. 13 (1963) 693–701.

[172]  J. G. Rosenstein, Linear Orderings, Pure and Applied Mathematics 98, Academic Press, Inc., New York-London, 1982.

[173]  J. G. Rosenstein, Recursive linear orderings, in M. Pouzet and D. Richard, eds., Orders: Description and Roles, North-Holland Math. Stud. 99, North-Holland, Amsterdam, 1984, 465–475.

[174]  G. E. Sacks, The recursively enumerable degrees are dense, Ann. Math. (2) 80 (1964) 300–312.

[175]  G. E. Sacks, Higher Recursion Theory, Perspectives in Mathematical Logic, Springer-Verlag, Berlin, 1990.

[176]  D. S. Scott, Algebras of sets binumerable in complete extensions of arithmetic, in Proceedings of the Symposium on Pure and Applied Mathematics, vol. V, American Mathematical Society, Providence, RI, 1962, 117–121.

[177]  D. Seetapun and T. A. Slaman, On the strength of Ramsey's Theorem, Notre Dame J. Formal Logic 36 (1995) 570–582.

[178]  J. Shinoda and T. A. Slaman, Recursive in a generic real, J. Symbolic Logic 65 (2000) 164–172.

[179]  J. R. Shoenfield, Degrees of formal systems, J. Symbolic Logic 23 (1958) 389–392.

[180]  J. R. Shoenfield, On degrees of unsolvability, Ann. of Math. (2) 69 (1959) 644–653.

[181]  R. A. Shore, Splitting an $\alpha$-recursively enumerable set, Trans. Amer. Math. Soc. 204 (1975) 65–77.

[182]  R. A. Shore, On the strength of Fraïssé's conjecture, in J. N. C. Crossley, J. Remmel, R. A. Shore and M. Sweedler, eds., Logical Methods: In Honor of Anil Nerode's Sixtieth Birthday, Progr. Comput. Sci. Appl. Logic 12, Birkhäuser, Boston, 1993, 782–813.

[183]  R. A. Shore, Reverse mathematics: the playground of logic, Bull. Symbolic Logic 16 (2010) 378–402.

[184]  R. A. Shore, Reverse mathematics, countable and uncountable: a computational approach, in [73] 150–163.

[185]  S. G. Simpson, Degrees of unsolvability: a survey of results, in J. Barwise, ed., Handbook of Mathematical Logic, North-Holland, Amsterdam, 1977, 1133–1142.

[186]  S. G. Simpson, Nonprovability of certain combinatorial properties of finite trees, in L. A. Harrington, M. D. Morley, A. Scedrov, and S. G. Simpson, eds., Harvey Friedman's Research on the Foundations of Mathematics, Stud. Logic Found. Math. 117, North-Holland, Amsterdam, 1985, 87–117.

[187]  S. G. Simpson, Partial realizations of Hilbert's program, J. Symbolic Logic 53 (1988) 349–363.

[188]  S. G. Simpson, Subsystems of Second Order Arithmetic, First edition, Perspectives in Mathematical Logic, Springer-Verlag, Berlin, 1999.

[189] S. G. Simpson, ed., Reverse Mathematics 2001, Lecture Notes in Logic 21, Association for Symbolic Logic, La Jolla, CA and A K Peters, Ltd., Wellesley, MA, 2005.

[190] S. G. Simpson, The Gödel hierarchy and reverse mathematics, in S. Feferman, C. Parsons, and S. G. Simpson, eds., Kurt Gödel: Essays for his Centennial, Lecture Notes in Logic 33, Association for Symbolic Logic, La Jolla, CA and Cambridge University Press, Cambridge, 2010, 109–127

[191] S. G. Simpson, Subsystems of Second Order Arithmetic, Second edition, Perspectives in Logic, Cambridge University Press, Cambridge and Association for Symbolic Logic, Poughkeepsie, NY, 2009.

[192] S. G. Simpson, Degrees of unsolvability, lecture notes, available at time of writing at http://www.math.psu.edu/simpson/notes/dou.pdf.

[193] S. G. Simpson and J. Rao, Reverse algebra, in [55], vol. 2, 1355–1372.

[194] T. A. Slaman, Recursion theory, presented as The Gödel Lecture, Annual Meeting of the Association for Symbolic Logic, Philadelphia, PA, 2001, slides available at time of writing at http://math.berkeley.edu/~slaman/talks/.

[195] T. A. Slaman, $\Sigma_n$-bounding and $\Delta_n$-induction. Proc. Amer. Math. Soc. 132 (2004) 2449–2456.

[196] R. I. Soare, Recursively Enumerable Sets and Degrees, Perspectives in Mathematical Logic, Springer-Verlag, Berlin, 1987.

[197] R. I. Soare, Computability Theory and Applications: The Art of Classical Computability, to be published by Springer.

[198] R. M. Solovay, Hyperarithmetically encodable sets, Trans. Amer. Math. Soc. 239 (1978) 99–122.

[199] E. Specker, Ramsey's Theorem does not hold in recursive set theory, in R. O. Gandy and C. E. M. Yates, eds., Logic Colloquium '69, Stud. Logic Found. Math., North-Holland, Amsterdam, 1971, 439–442.

[200] C. Spector, On degrees of recursive unsolvability, Ann. of Math. (2) 64 (1956) 581–592.

[201] H. Towsner, Hindman's Theorem: an ultrafilter argument in second order arithmetic, J. Symbolic Logic 76 (2011) 353–360.

[202] H. Towsner, A simple proof and some difficult examples for Hindman's Theorem, Notre Dame J. Formal Logic, 53 (2012) 53–65.

[203] H. Towsner, Ultrafilters in reverse mathematics, J. Math. Log. 14 (2014) 1450001, 11 pp.

[204] H. Wang, Popular Lectures on Mathematical Logic, Revised reprint of the 1981 second edition, Dover Publications, Inc., New York, 1993.

[205] W. Wang, Rainbow Ramsey Theorem for triples is strictly weaker than the Arithmetic Comprehension Axiom, J. Symbolic Logic 78 (2013) 824–837.

[206] W. Wang, Cohesive sets and rainbows, Ann. Pure Appl. Logic 165 (2014) 389–408.

[207] W. Wang, Some logically weak Ramseyan theorems, Adv. Math. 261 (2014) 1–25.

[208] K. Weihrauch, The degrees of discontinuity of some translators between representations of the real numbers, Technical Report TR-92-050, Interna-

214

*Slicing the Truth*

tional Computer Science Institute, Berkeley, 1992.

[209]  K. Weihrauch, The TTE-interpretation of three hierarchies of omniscience principles, Informatik Berichte FernUniversität Hagen 130, Hagen, 1992.

[210]  C. E. M. Yates, Arithmetical sets and retracing functions, Z. Math. Logik Grundlag. Math. 13 (1967) 193–204.

[211]  X. Yu and S. G. Simpson, Measure theory and weak König's lemma, Arch. Math. Logic 30 (1990) 171–180.

Printed in the United States
by Bookmasters

Printed in the United States
By Bookmasters